DESCRIPTIVE STATISTICS WITH R

César Pérez López

CONTENTS

Chapter 1

R'S WORKING ENVIRONMENT FOR STATISTICAL ANALYSIS

1.1 THE SOFTWARE R

R is a language and programming environment for statistical and graphical analysis. It is a free software project, the result of the GNU implementation of the classic scientific language S-Plus. R and S-Plus are probably the two most used languages in research by the statistical community, being also very popular in the field of biomedical research, bioinformatics and financial mathematics. The possibility of loading different libraries or packages for specific calculation or graphical purposes contributes to this. R is distributed under the GNU GPL license and is available for Windows, Macintosh, Unix and GNU/Linux operating systems.

R provides a wide range of statistical tools (linear and non-linear models, statistical tests, time series analysis, classification and clustering algorithms, etc.) and graphics. Like S Plus, it is a programming language, which allows users to extend it by defining their own functions. In fact, many of the functions of R are written in R itself, although for computationally demanding algorithms it is possible to develop libraries in C, C + + or Fortran that are loaded dynamically. More advanced users can also manipulate R objects directly from code developed in C. R can also be extended through packages developed by your user community.

R is part of a collaborative and open project. Your users can publish packages that extend their basic configuration. Given the enormous number of new packages, they have been organized into views (or themes), which allow them to be grouped according to their nature and function. For example, there are groups of packages related to Bayesian statistics, econometrics, time series, etc.

R, like S plus, can be considered an object oriented programming language. In addition, R can be integrated with different databases and there are libraries that facilitate its use from programming languages interpreted as Perl and Python. R also has its own format for documentation based on LaTeX.

Another feature of R is its graphic capability, which allows for the generation of high quality graphics.

R can also be used as a numerical calculation tool, where it can be as effective as other specific tools such as GNU Octave and its commercial equivalent, MATLAB.

1.2 WORKING ENVIRONMENT OF R

We already know that in R you can work in command mode or in interactive window mode. The command mode is the default mode (*R Console*).

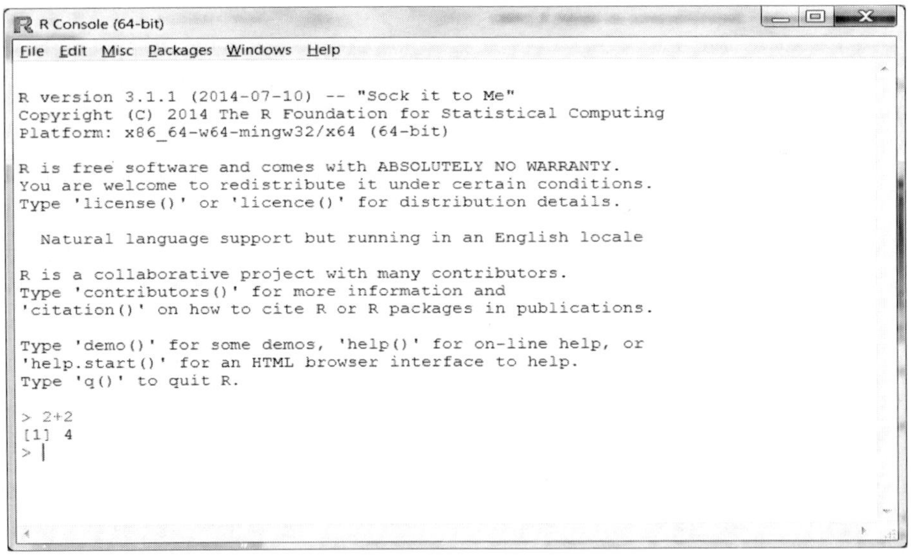

The *R Commander* interactive window mode is the one obtained after installing and reading the *Rcmdir* package. R Commander is displayed with 3 windows for instructions (*R Script)*, results (*Output*) and error messages (*Messages*).

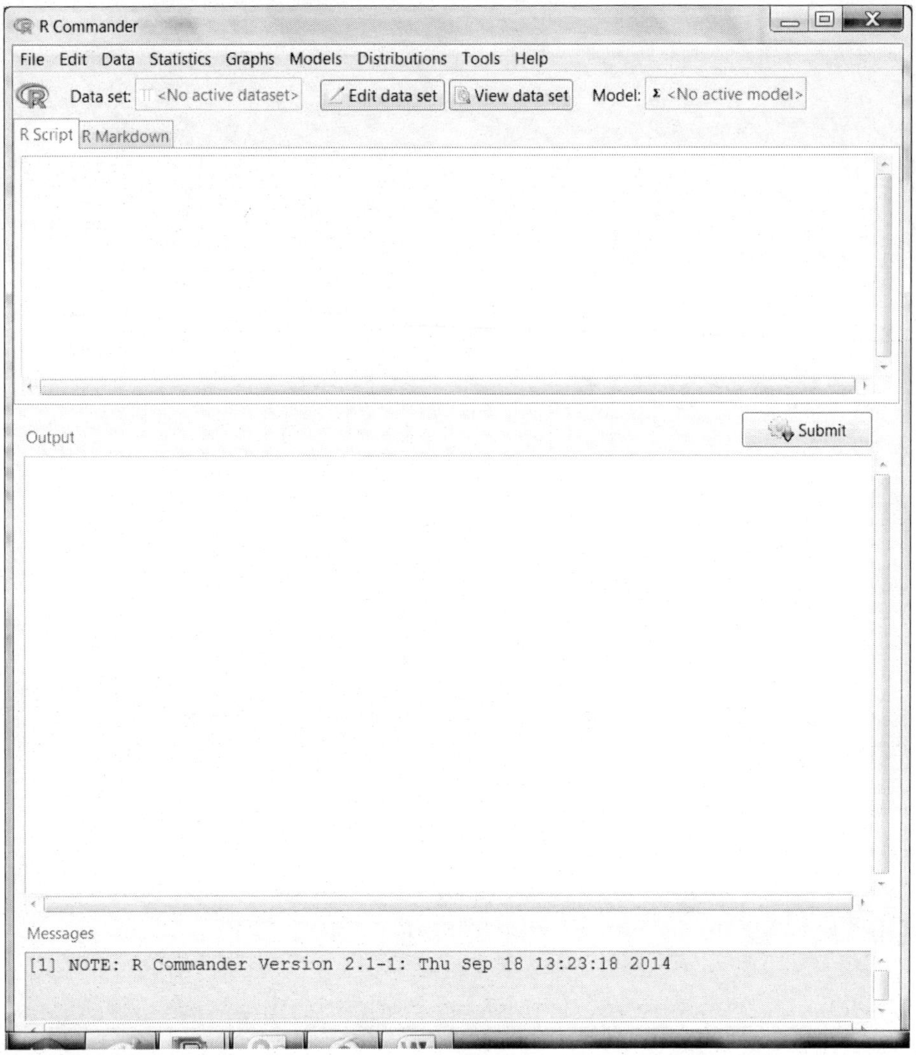

1.3 INSTALLATION OF ADDITIONAL STATISTICAL PACKAGES

So far we have installed only the basic R software. But there are many additional packages that extend the program in many facets of work. To install an additional package, in the *Packages* menu, click on *Install package(s)*. Then choose mirror in the *CRAN mirror* window (if possible the closest one), click *Ok* and in the *Packages* screen choose the packages to be installed. In our case, we will initially choose *fBasics* (basic statistics) and *Rcmdr (*interactive R command window). Clicking Ok installs the packages.

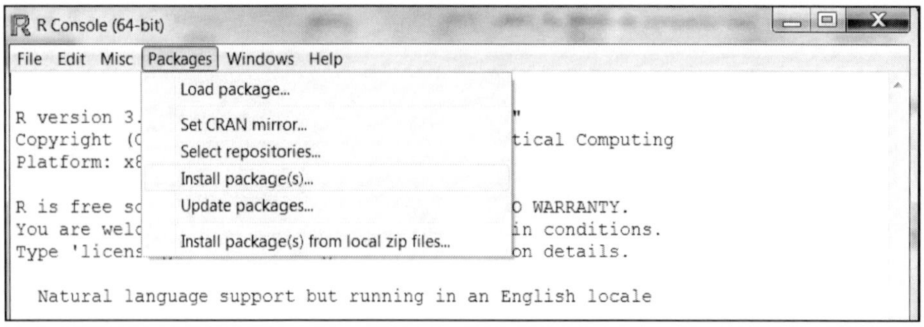

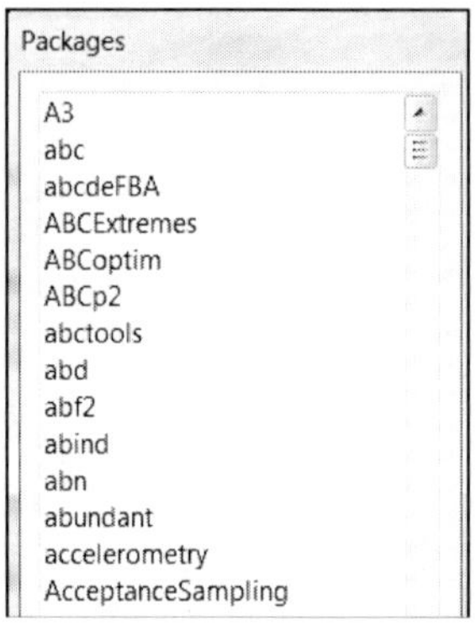

In this way, additional program packages are installed to the base package. This process of installing packages is done only the first time a package is installed. Once installed, the option *Load package is* chosen in the menu Packages and in the list of resulting packages *fBasics* and *Rcmdr* are chosen again, to be used. In this way, the available packages that you have installed and want to use in an R session are placed as active. The packages to be used must be loaded with *Load package* when the R session is started.

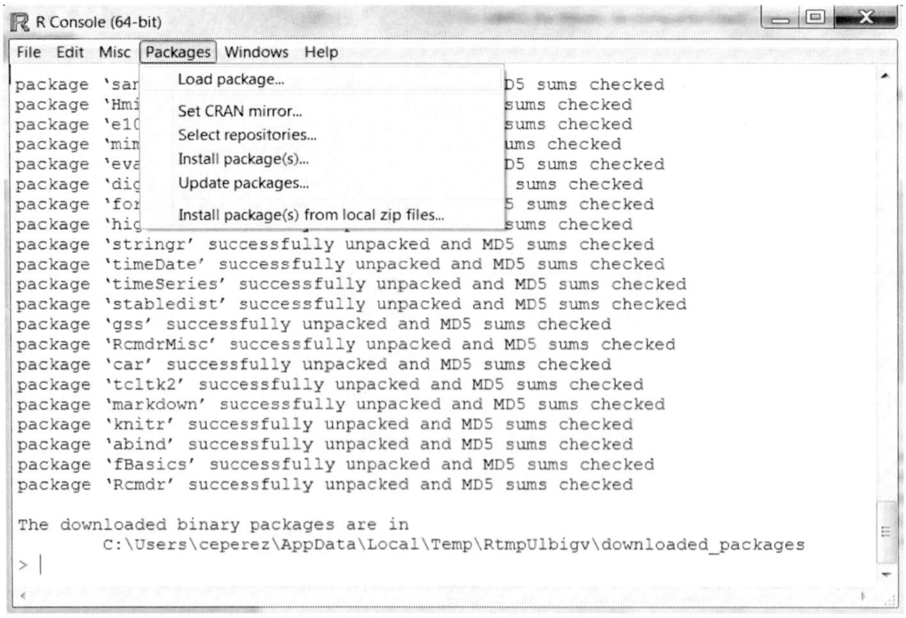

1.4 R STUDIO

R Studio is an R environment that complements *R Console* and *Rcomander*. It is free software that can be obtained at the official website of the program http://www.rstudio.com/products/rstudio/download/

R Studio's working environment is similar to R Console and R Comander. On the left of the screen you can see the command area which is similar in all three environments. The menu bar is the element that differentiates the three environments. You can see that the simplest is R Console. R Comander has a richer menu bar and is very similar to R Studio. As we have already seen, R commander has three windows whose purposes are to enter commands, see the output and see comments on the execution of sentences and programs. For its part, the R Studio environment also has three windows. The left window is the command window. The top right window allows you to see the environment variables and their value as you work (*Environment) and* also allows you to see the history of executed commands (*History*). The bottom window allows you to see the working directory.

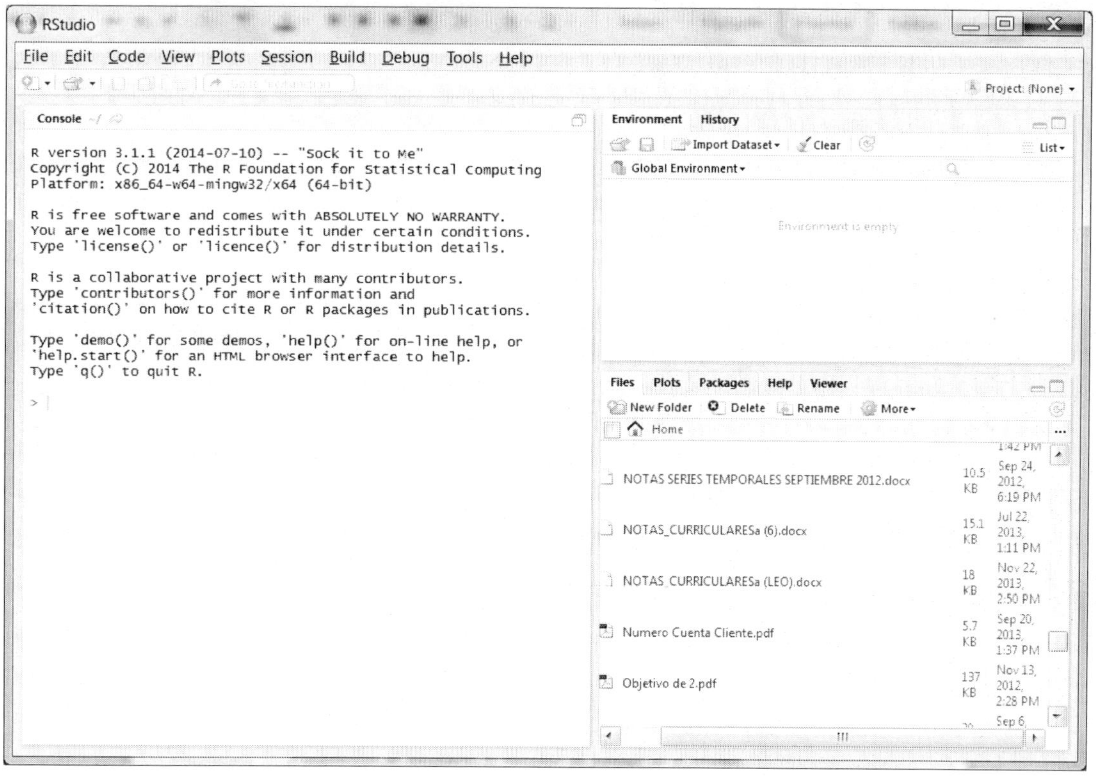

The *Install Packages* option under the *Tools* menu allows you to install additional packages in the R Studio environment. All you need to do is enter your name in the *Packages* field. You can also choose the source of the packages and their destination on disk.

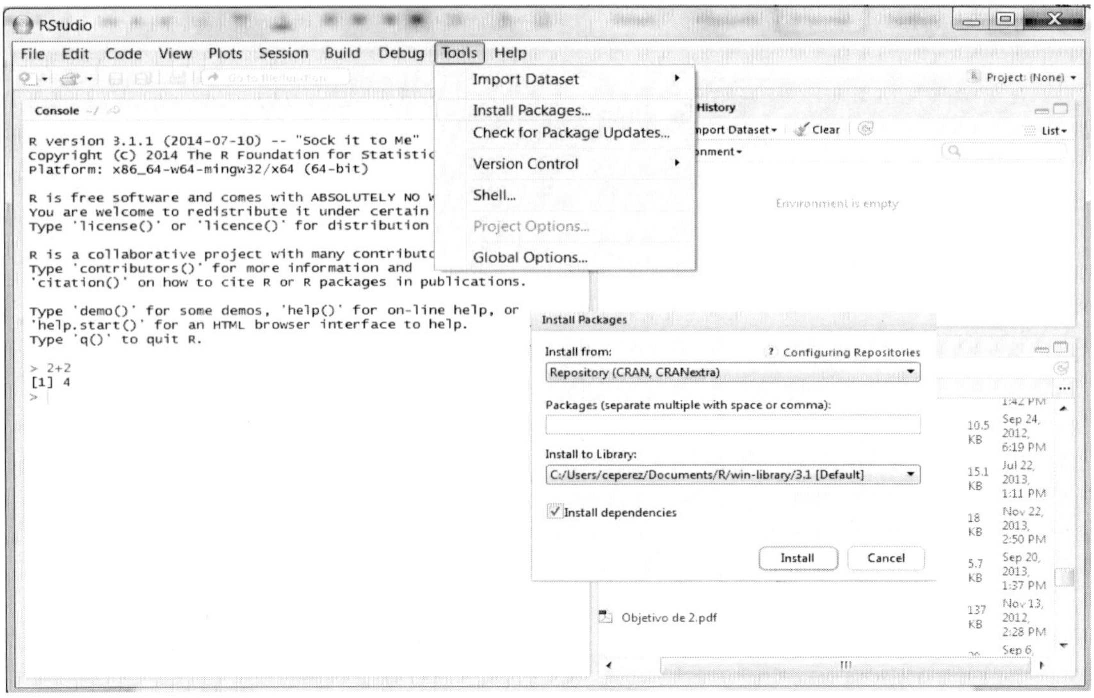

Once a package is installed, the *Packages:* tab in the lower right part of the R Studio window allows you to view the list of available packages (they are marked in the list) and those already installed. To enable any of them, simply check it

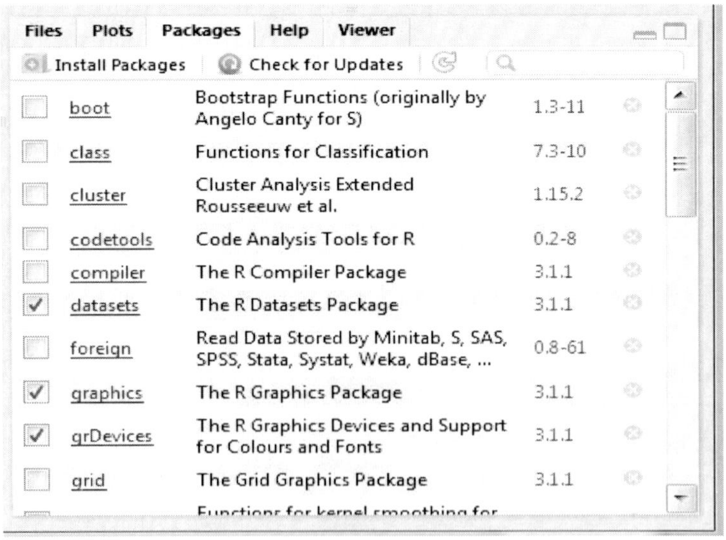

To use a new package, for example *cluster,* just mark it in the packages list and automatically in the command window the command to read the package is executed. If you unmark a packet in the packages list, the syntax in the command window is automatically generated, which disables the use of the packet.

1.5 IMPORTING AND EXPORTING DATA

In R you can read data from external files with formats suitable for most statistical packages. It is possible to read ASCII, SPSS, STATA, etc. files. However, R has its own file format with the RData extension. It is also possible to write files in R format.

1.5.1 R-format files

The menu bar options on the Environment screen of the R Studio environment allow you to perform data import and export tasks.

The icon allows you to save the current information in the command window to a file in R format (extension RData) that can later be loaded into memory with the icon when you start another session. The following figures illustrate the task of saving the working environment information into the file *commandData.RData* and its subsequent reading.

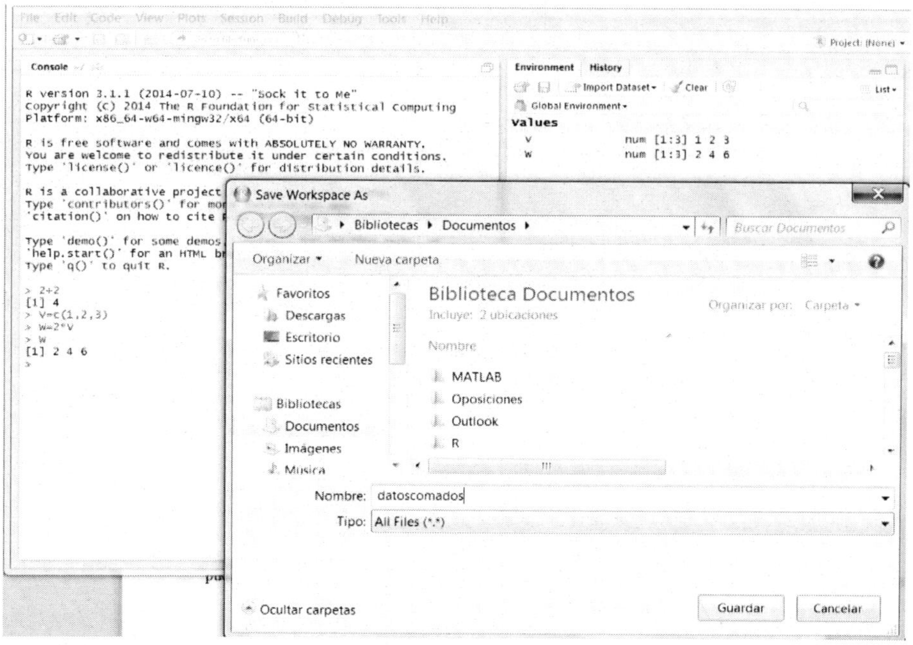

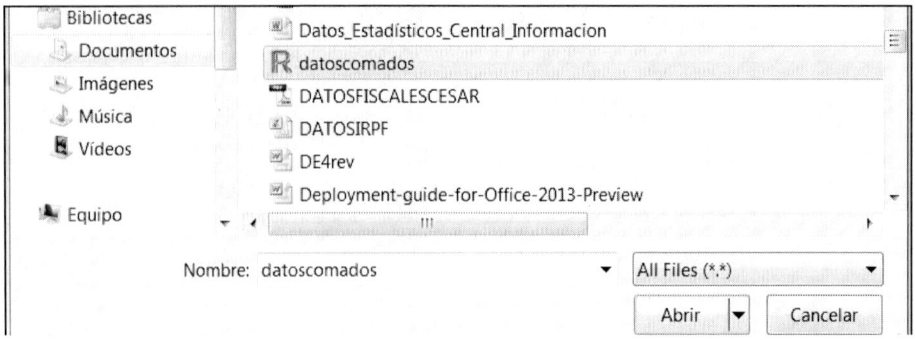

In the command window you can see the syntax of R equivalent to the tasks of placing a working directory, saving the workspace to an R file and then reading that saved information from the workspace.

```
> setwd("C:/R/DATA")
> save.image("C:/R/DATA/datoscomandos.RData")
> load("C:/R/DATA/datoscomandos.RData")
```

1.5.2 Files in ASCII format

The *Import Dataset* option from the *Environment* menu allows you to import external files in ascii format and web files.

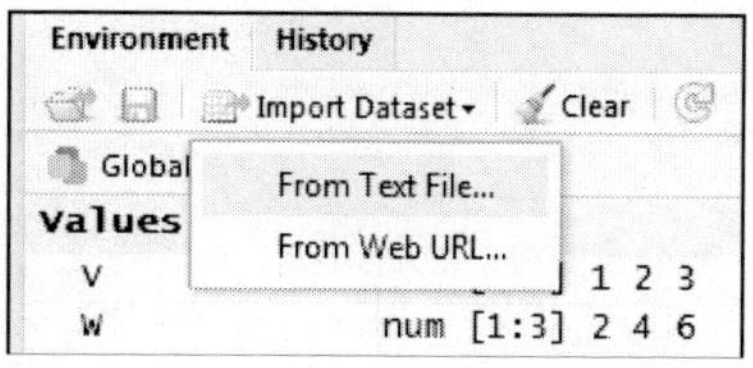

The *From Text File* option in *Import Dataset* allows you to import external ascii files. When importing, R presents an image of the file content and the options to use variable names in the first line of the file (*Heading), to* choose a specific character separator (tab, white space, comma and semicolon), to represent decimals with periods or commas and to enter characters between single or double quotes.

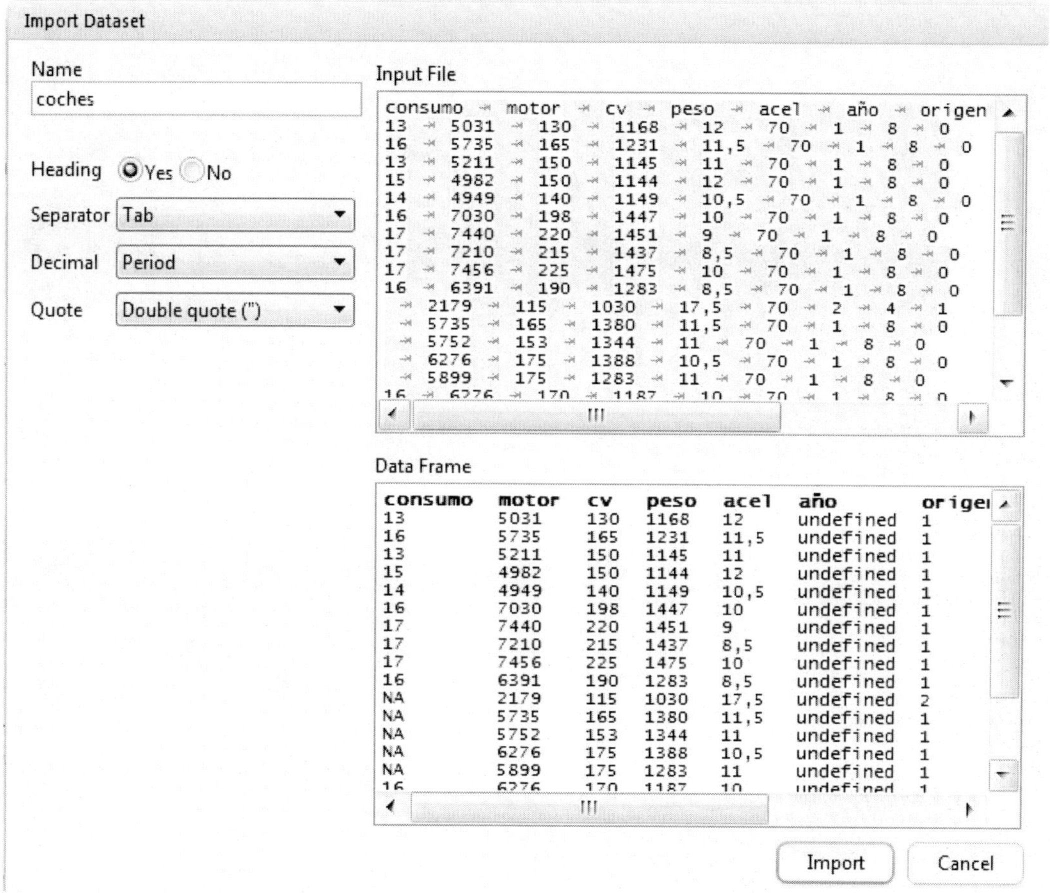

When you press Import the ascii file is imported into the R workspace and it is already possible to work in R with your information.

In the command window you can see the syntax used for i9mporting an ASCII file and displaying its contents in the dwe workspace.

```
> cars <- read.delim("C:/R/DATA/coches.txt", dec=",")
> View(cars)
```

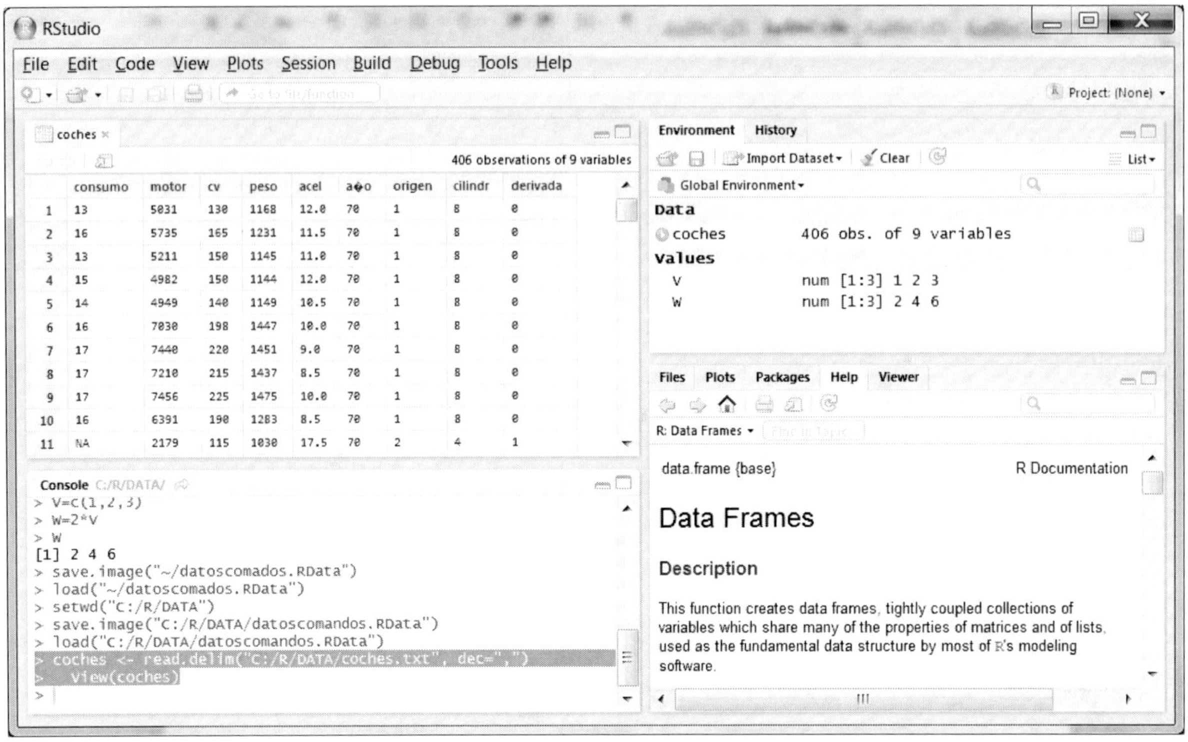

The import of external files in ASCII format can also be done from the option *Tools → Import Data Set from* the general menu of R Studio.

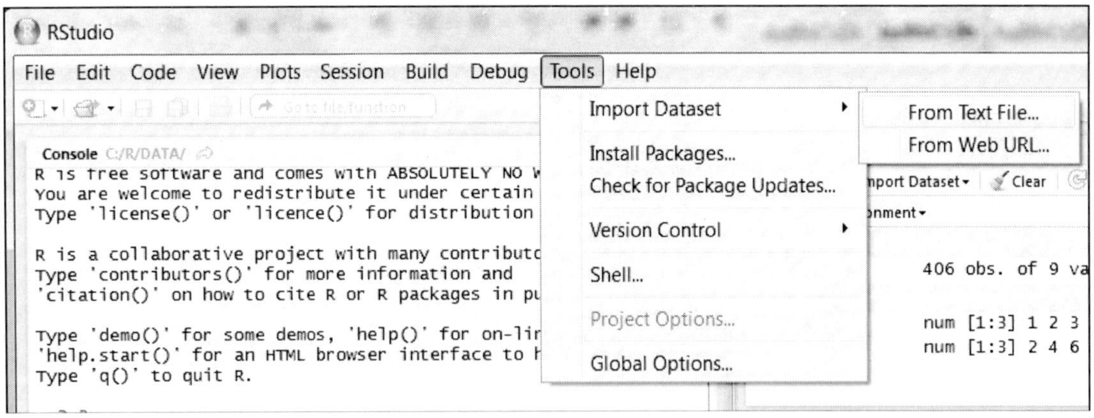

1.5.3 Files in SPSS, SAS, STATA, S, MINITAB, SYSTAT, OCTAVE and other formats

R can read files in SPSS, SAS, STATA, S, MINITAB, SYSTAT, OCTAVE and other formats through the *foreing* library. You start by enabling the library through the ioption packages in the R Studio environment (or installing it through the *Install Packages* suboption of the *Tools* option in the R Studio menu bar, if it is not installed).

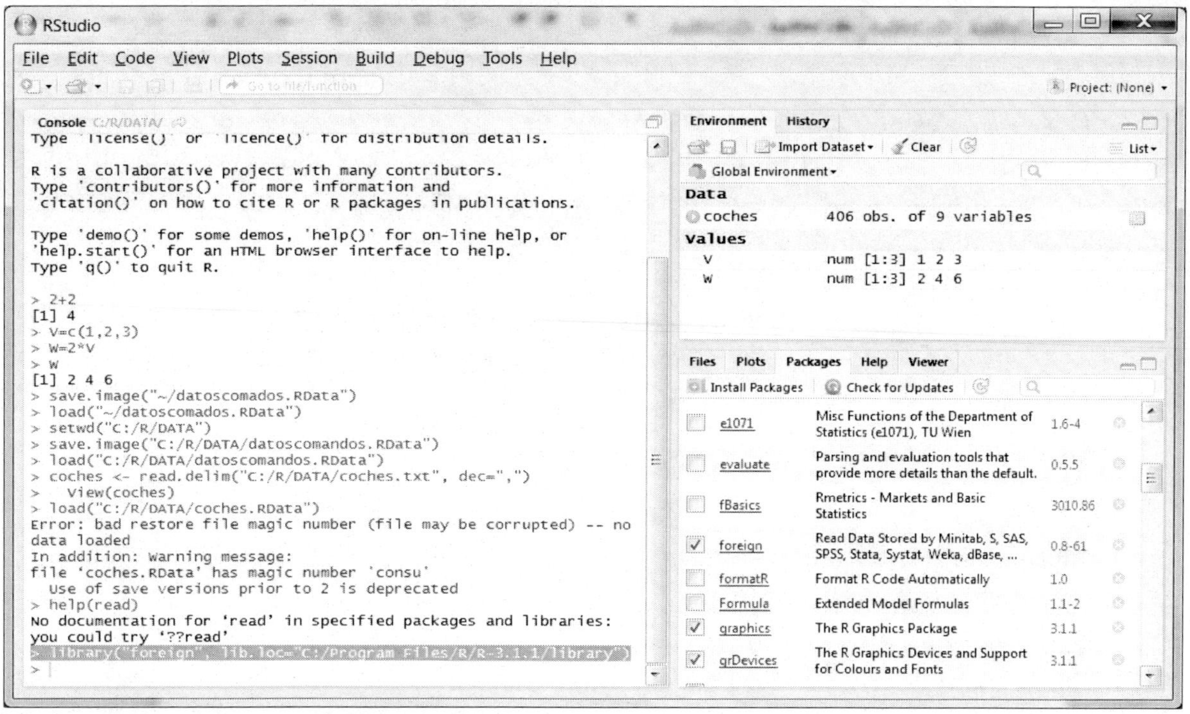

In the previous figure it is observed that the syntax equivalent to the reading of the library is the following:

```
> library("foreign", lib.loc="C:/Program Files/R/R-3.1.1/libr
ary")
```

Then you can use the *read.spss* function to read the file.

```
> read.spss("c:/R/DATA/coches.sav")
```

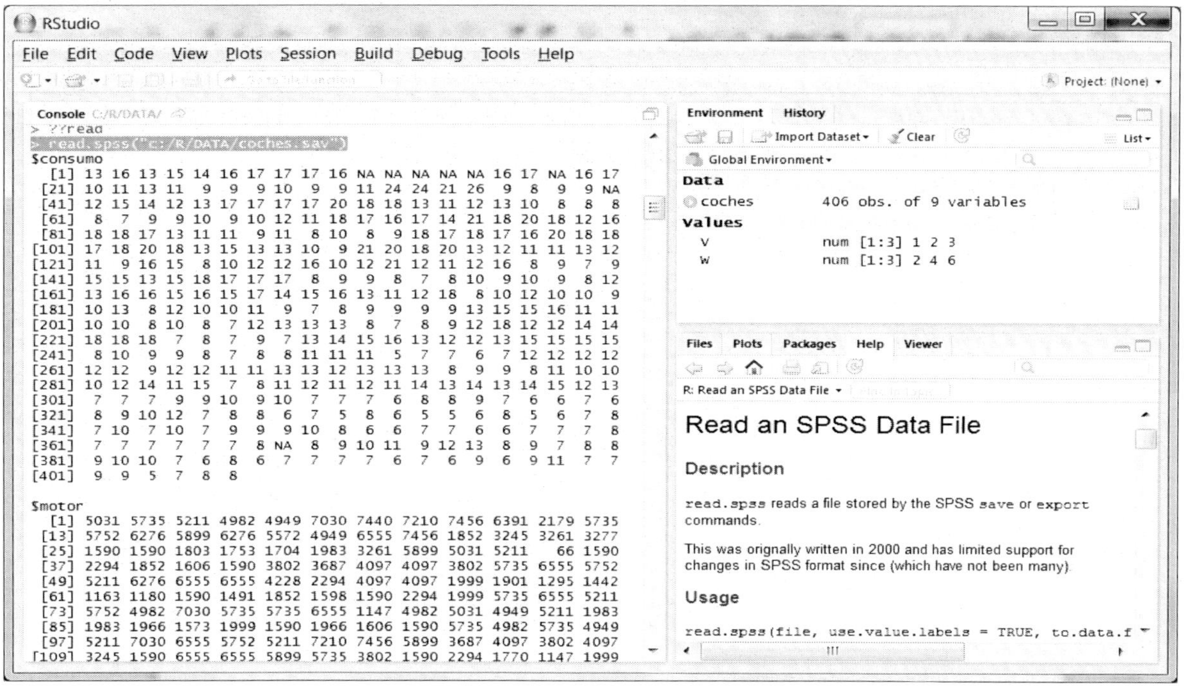

In the same way you can read files of the formats SAS, STATA, S, SYSTAT, MINITAB, SAS Transport and OCTAVE using respectively the functions *read.sas7badt, read.dta, read.S, read.systat, read.mto, read.xport* and read.*octave and* read.dta of the *foreing* library. The most general syntax for all these functions is *read.format ("file")*. The syntax of the read command used here is the simplest possible. An extended syntax could be as follows:

```
read.spss(file, use.value.labels = TRUE, to.data.frame =
FALSE, max.value.labels = Inf, trim.factor.names = FALSE,
trim_values = TRUE, reencode = NA, use.missings =
to.data.frame)
```

The option *use.value.labels = TRUE* allows you to convert the values of the labels into R factors with their corresponding levels.

The option *to.data.frame = TRUE* allows you to convert the data file into a *dataframe* structure, which will be explained later.

The option *max.value.labels = Inf* allows to convert to factors all the variables independently of their labels, since their values are not limited.

The option *trim.factor.names = TRUE* allows to compress the spaces in the factor levels.

The option *trim_values = TRUE* allows to compress the spaces of the values of the labels

The option *reencode = NA* allows to recode as NA the missing information.

The option *use.missings = to.data.frame* allows to use as deasparecidos the elements that have the value defined in *reencode* (usually NA).

The syntax is similar for all *read.format(file)* commands no matter what the format is.

Chapter 2

BASIC ELEMENTS IN THE STATISTICS WORK: NUMBERS, OPERATORS, VECTORS, MATRICES AND FUNCTIONS

2.1 NUMBERS AND ARITHMETIC OPERATIONS

The numerical work in R is very broad. You can use integers, real and complex, which in turn can be function arguments giving rise to the integer functions, real and complex. Therefore, the complex variable is a field treatable in R. The arithmetic operations in R are defined according to standard mathematical conventions. The following table presents the syntax of the basic arithmetic operations:

x + y	*Sum*
	`> 4+8`
	`[1] 12`

x - y	*Difference*
	`> 4-8`
	`[1] -4`

x * y	*Product*
	`> 4*8`
	`[1] 32`

x/y	*Division*
	`> 4/8`
	`[1] 0.5`

x ^y or xy**	*Power*
	`> 4**8`
	`[1] 65536`

2.2 INTEGERS AND INTEGER FUNCTIONS

R works with integers and integer functions accurately. Regardless of the format in which the results are presented, the calculations are accurate. However, the *vpa* command allows you to obtain exact outputs with the required precision.

As for the functions with integer variable, the most important ones that R contemplates are the following (the expressions between quotes are in string format):

sign(n)	*Sign of n (1 if n>0 , -1 if n<0, n real)*
	`> sign(-8)`
	`[1] -1`

max(n1,n2)	*Maximum of the numbers n1 and n2*
	`> max(17,12)`
	`[1] 17`

min(n1,n2)	*Minimum of the numbers n1 and n2*
	`> min(17,12)`
	`[1] 12`

factorial(n)	*Factorial of n (n(n-1)(n-2)...1)*
	`> factorial(4)`
	`[1] 24`

choose(n,k)	*Combinatorial number n over k*
	`> choose(5,2)`
	`[1] 10`

2.3 REAL NUMBERS AND REAL VARIABLE FUNCTIONS

Real numbers are the disjointed union of rational numbers and irrational numbers. A rational number is of the form *p/q,* where *p is* an integer and *q* another integer. The way R treats rational numbers is different from most calculators. Rational numbers are quotients of integers, and R can also work with them in an exact way, so that the result of expressions involving rational numbers is always another rational or integer number. R returns decimal approximations of the results. We consider the following example of operations between rational numbers:

```
> 1/6+1/5-2/10
[1] 0.1666667
```

R also works with irrational numbers by representing the results as accurately as it can. Here is an example.

```
> sqrt(235)
[1] 15.32971
```

There are very typical real constants that R represents as follows:

pi	*Number π= 3,1415926*
	`> 2*pi`
	`[1] 6.283185`

exp(1)	*Number e = 2,7182818*
	`> exp(1)`
	`[1] 2.718282`

inf	*Infinity (e.g. 1/0)*
	`> 1/0`
	`1] Inf`

NaN	*Indetermination (e.g. 0/0)*
	`> 0/0`
	`1] NaN`

NA	*Missing value*
NULL	*Null object*

R has a range of predefined real variable functions, which of course will be valid for rational, irrational and integer variables. In the following paragraphs the most important ones are presented.

2.3.1 Trigonometric functions

Below is a table with the trigonometric functions and their inverses that incorporates R illustrated with examples.

Function	***Reverse***

sin(x) *Sine*
`> sin(pi/2)`
`[1] 1`
	asin(x) *Sine arc*
	`> asin(1)`
	`[1] 1.570796`

cos(x) *Cosine*
`> cos(pi)`
`[1] -1`
	acos(x) *Cosine arc*
	`> acos(-1)`
	`[1] 3.141593`

tan(x) *Tangent*
`> tan(pi/4)`
`[1] 1`
	atan(x) and atan2(x)
	`> atan(1)`
	`[1] 0.7853982`

2.3.2 Hyperbolic functions

Below is a table with the hyperbolic functions and their inverses that incorporates R illustrated with examples.

Function	*Reverse*

sinh(x) *Hyperbolic sine*
`> sinh(2)`
`[1] 3.62686`
	asinh(x) *Hyperbolic sine arc*
	`> asinh(3.6269)`
	`[1] 2.000011`

cosh(x) *Hyperbolic cosine*
`> cosh(3)`
`[1] 10.06766`
	acosh(x) *Hyperbolic cosine arc*
	`> acosh(10.0677)`
	`[1] 3.000004`

tanh(x) *Hyperbolic Tangent*
`> tanh(1)`
`[1] 0.7615942`
	atanh(x) *Hyperbolic arc tangent*
	`> atanh(0.7616)`
	`[1] 1.000014`

2.3.3 Exponential and logarithmic functions

Below is a table with the exponential and logarithmic functions that R incorporates illustrated with examples.

Function	*Meaning*

exp(x)	*Exponential function in base e (e^x)*
	`> exp(log(7))`
	`[1] 7`

log(x)	*Logarithm function in base e of x*
	`> log(exp(7))`
	`[1] 7`

log10(x)	*Logarithm function in base 10 of x*
	`> log10(1000)`
	`[1] 3`

log2(x)	*Logarithm function in base 2 of x*
	`> log2(2^8)`
	`[1] 8`

sqrt(x)	*Square root of x function*
	`> sqrt(2^8)`
	`[1] 16`

2.3.4 Specific functions of numerical variable

R incorporates a group of functions of numerical variable whose purpose is to obtain approximate results of operations and roundings, work with signs, etc. Among these functions we have the following:

Function	*Meaning*

abs(x)	*Absolute value of real x*
	`> abs(-8)`
	`[1] 8`

floor(x)	*The largest integer less than or equal to the real x*
	`> floor(-23.557)`
	`[1] -24`

ceiling(x)	*The smallest integer greater than or equal to the real x*
	`> ceiling(-23.557)`
	`[1] -23`

round(x)	*The integer closest to the real x*
	`> round(-23.557)`
	`[1] -24`

round(x,n)	*Round off elements from x to n decimals*

trunc(x)	*Removes the decimal part of the real x*
	`> trunc(-23,557)`
	`[1] -23`

sign(x)	*Sign of the real x (1 if x>0 , -1 if x<0)*
	`> sign(-23.557)`
	`[1] -1`

2.4 SIMPLE VARIABLES

MATLAB does not require any type of command to declare variables. It simply creates the variable by directly assigning its value. For example:

```
> V=3
> V
[1] 3
```

The variable *V* will be worth 3 as long as its value is not changed by a new assignment. Once declared the variable we can use it in calculations.

```
> V^3
[1] 27

        27
```

```
> V+5
[1] 8
```

The value assigned to a variable is permanent, until it is not changed expressly or until it leaves the current session of R.

If we write now:

```
> V=3+7
> V
```

```
[1] 10
```

the variable *V becomes* worth 10 from this moment, as observed in the following calculation:

```
> V^4
[1] 10000
```

Variable names begin with a letter followed by any number of letters, digits, or underscores. It is also very important to note that R is case-sensitive; therefore, a variable with an uppercase letter is different from the same variable with a lowercase letter.

2.5 VECTOR VARIABLES

To represent a vector of *n* elements, a variable can be defined in R as follows:

V = c(v1, v2, v3, ,vn.......)

When most R commands and functions are applied to a vector variable the result is the application of the command or function to each element of the vector:

```
> vector1=c(1,4,9,2.25,1/4)
> vector1
[1] 1.00 4.00 9.00 2.25 0.25

> sqrt(vector1)
[1] 1.0 2.0 3.0 1.5 0.5
```

There are also specific ways of defining vector variables, among which are the following:

variable=a:b	*It defines the vector whose first and last elements are a and b, respectively, and the intermediate elements are differentiated into a unit* > V1=1:6 > V1 [1] 1 2 3 4 5 6
variable=seq(a:b)	*It defines the vector whose first and last elements are a and b, respectively, and the intermediate elements are differentiated into a unit (sequence)* > V1=seq(1:6) > V1 [1] 1 2 3 4 5 6

<table>
<tr><td>variable=seq(a,b, by=s)</td><td>Defines the vector whose first and last elements are a and b, and the intermediate elements differ in the amount s specified by the increment

<code>> V2=seq(2,8,by=2)</code>
<code>> V2</code>
<code>[1] 2 4 6 8</code></td></tr>
<tr><td>variable=seq(along=vector)</td><td>Defines the vector of consecutive integers whose first and last elements are 1 and the length of the vector

<code>> seq(along=V1)</code>
<code>[1] 1 2 3 4 5 6</code></td></tr>
<tr><td>variable=numeric(n)</td><td>Create the vector with n null elements

<code>> V3=numeric(10)</code>
<code>> V3</code>
<code>[1] 0 0 0 0 0 0 0 0 0 0</code></td></tr>
<tr><td>variable=rep(x,n)</td><td>Repeat n times the value x or the vector x

<code>> V4=rep(1,6)</code>
<code>> V4</code>
<code>[1] 1 1 1 1 1 1</code>

<code>> v5=rep(c(1,2,3),3)</code>
<code>> v5</code>
<code>[1] 1 2 3 1 2 3 1 2 3</code></td></tr>
<tr><td>variable=
list(object1, ..., object)</td><td>Creates a list with the specified objects

<code>> V6=list(a=c(1,2),b="hola",c=3)</code>
<code>> V6</code>
<code>$a</code>
<code>[1] 1 2</code>

<code>$b</code>
<code>1] "hello"</code>

<code>$c</code>
<code>[1] 3</code></td></tr>
</table>

2.5.1 Vector variable elements

R allows the selection of elements of vector variables using the following commands:

x[n]	*Returns the umpteenth element of vector x*
	
	`> X=2:8`
	`> X`
	`[1] 2 3 4 5 6 7 8`
	`> X[3]`
	`[1] 4`

x[-n]	*Returns all values of vector x except the nth*
	
	`> X[-3]`
	`[1] 2 3 5 6 7 8`

x[1:n]	*Returns the first n elements of vector x*
	
	`> X[1:3]`
	`[1] 2 3 4`

x[a:b]	*Returns the elements of vector x between the a- and b-vector*
	
	`> X[2:3]`
	`[1] 3 4`

x [-(1:n)]	*Returns the elements of vector x from n+1 to the last*
	
	`> X[-(1:3)]`
	`[1] 5 6 7 8`

x [-(a:b)]	*Returns the elements of vector x that are not between the a- and b-vector*
	
	`> X[-(2:3)]`
	`[1] 2 5 6 7 8`

x[c(a,b,c)]	*Returns the elements of vector x located in the a-th, b-th and c-th places*
	
	`> X [c(3,5,6)]`
	`[1] 4 6 7`

x[x>a]	*Returns the elements of vector x that are larger than the value a*
	
	`> X [X>4]`
	`[1] 5 6 7 8`

x [x>a & x<b]	*Returns the elements of vector x that are larger than the value a and smaller than the value b*

	`> X [X>4 & X<7]` `[1] 5 6`

x["name"]	*Returns element of vector or list x named "name*
	
	`> X=list(c(1,2,3), "Jhon",C=7)`
	`> X["C"]`
	`$C`
	`[1] 7`

x[a]	*Returns the a-item in the x-list*
	
	`> X=list(c(1,2,3), "Jhon", 7)`
	`> X`
	`[[1]]`
	`[1] 1 2 3`
	
	`[[2]]`
	`1] "John"`
	
	`[[3]]`
	`[1] 7`
	
	`> X[2]`
	`[[1]]`
	`1] "John"`

2.5.2 Vector variable functions

The real single-variable functions are expandable to vectorial vraibale. To apply a single-variable function to a vector variable, just apply the function to each element of the vector. Below are examples of trigonometric and inverse trigonometric functions.

Function	***Reverse***

sin(V) *Sine*
 	
`> v=c(0, pi/2, pi,3*pi/2,2*pi)`
`> round(sin(v))`
`[1] 0  1 0 -1 0`
	asin(V) *Sine arch*
	
	`> w=c(0,1/2,1)`
	`> asin(w)`
	`[1] 0.0000000 0.5235988 1`
	`.5707963`

cos(V) *Cosine*
 	
`> round(cos(v))`
`[1] 1  0 -1 0  1`
	acos(V) *Cosine arc*
	
	`> acos(w)`
	`[1] 1.570796 1.047198 0.0`
	`00000`

tan(V) *Tangent* `> round(tan(v))`	**atan(V) and atan2(V)** `> atan(w)`

1] 0.000000e+00 1.633124e+16 0 .000000e+00 5.443746e+15 0.000 000e+00	[1] 0.0000000 0.4636476 0 .7853982

sinh(V) *Hyperbolic sine*
 	
`> sinh(w)`
`[1] 0.0000000 0.5210953 1.1752 012`
	asinh(V) *Hyperbolic sine arc*
	
	`> asinh(w)`
	`[1] 0.0000000 0.4812118 0 .8813736`

cosh(V) *Hyperbolic cosine*
 	
`> cosh(w)`
`[1] 1.000000 1.127626 1.543081`
	acosh(V) *Hyperbolic cosine arc*
	
	`> acosh(w)`
	`1] NaN NaN 0`

tanh(V) *Hyperbolic Tangent*
 	
`> tanh(w)`
`[1] 0.0000000 0.4621172 0.7615 942`
	atanh(V) *Hyperbolic arc tangent*
	
	`> atanh(w)`
	`1] 0.0000000 0.5493061 Inf`

Below are examples of exponential, logarithmic, and other vector-variable functions:

Function	***Meaning***

exp(V)	*Exponential function in base e of the elements of vector V*
	
	`> exp(log(c(1,2,3))`
	`[1] 1 2 3`

log(V)	*Logarithm function on the basis of e of the elements of vector V*
	
	`> log(exp(c(1,2,3))`
	`[1] 1 2 3`

log10(V)	*Logarithm function on the basis of 10 of the elements of vector V*
	
	`> log10((c(10,100,1000)))`
	`[1] 1 2 3`

log2(V)	*Logarithm function on base 2 of the elements of vector V*
	
	`> log2(c(2,4,8))`
	`[1] 1 2 3`

sqrt(V)	*Square root function of Vector V elements*
	
	`> sqrt(c(2,4,8))`
	`[1] 1.414214 2.000000 2.828427`

floor(V)	*The largest integer less than or equal to each element of vector V*
	`> floor(sqrt(c(2,4,8))`
	`[1] 1 2 2`

ceiling(V)	*The smallest integer greater than or equal to each element of vector V*
	`> ceiling(sqrt(c(2,4,8))`
	`[1] 2 2 3`

round(V)	*The integer closest to each element of V*
	`> round(sqrt(c(2,4,8))`
	`[1] 1 2 3`

trunc(V)	*Removes the decimal part of each V*
	`> trunc(sqrt(c(2,4,8))`
	`[1] 1 2 2`

abs(V)	*Absolute value of each V element*
	`> abs(c(-1,0,1,-6,2))`
	`[1] 1 0 1 6 2`

sign(V)	*Sign of each V element (1 if Vi>0 and -1 if Vi<0)*
	`> sign(c(-1,0,1,-6,2))`
	`[1] -1 0  1 -1 1`

factorial(V)	*Factorial of each V element*
	`> factorial(c(1,2,3,4,5,6))`
	`[1]    1    2    6  24 120 720`

length(V)	*Number of elements of the vector V*
	`> length(c(1,2,3,4,5,6))`
	`[1] 6`

sum(V)	*Sum of the elements of the vector V*
	`> sum(c(3,8,9,15))`
	`[1] 35`

product(V)	*Product of Vector V elements*
	`> prod(c(3,8,9,15))`
	`[1] 3240`

2.5.3 Factors

A factor is a vector of character strings used to represent categorical data, which may also include the different levels of the categorical variable. Factors are created with the *factor* command. The following syntax creates a factor from the elements of a vector:

```
> data = c(1,2,2,3,1,2,3,3,1,2,3,3,1)
> fdata = factor(data)
> fdata
 [1] 1 2 2 3 1 2 3 3 1 2 3 3 1
Levels: 1 2 3
```

It can be seen that the above factor has three levels. It is possible to change the values of the levels of a factor with the command *labels*. For example, with the following syntax we can change the levels of the previous factor to Roman numerals:

```
> rate = factor(data,labels=c("I","II","III"))
> rat
 1] I    II II III I    II III III I    II III III I
Levels: I II III
```

We could also have changed the levels of a factor with the *levels* command, through the following syntax:

```
> levels(fdata) = c('I','II','III')
> fdata
 1] I    II II III I    II III III I    II III III I
Levels: I II III
```

The *levels* command can be used as an argument to the factor command.

```
> factor(1:3,labels=c("A","B","C"))
1] A B C
Levels: A B C
```

The *exclude* command allows you to exclude levels of a custom factor.

```
> factor(1:5,exclude=4)
1] 1     2     3 <NA>     5
Levels: 1 2 3 5
```

The command ordered allows you to specify whether the factor values are ordered

```
> fert = c(10,20,20,50,10,20,10,50,20)
> fert = factor(fert,levels=c(10,20,50),ordered=TRUE)
> fert
[1] 10 20 20 50 10 20 10 50 20
Levels: 10 < 20 < 50
```

2.6 MATRIX VARIABLES

To represent an array of *n* elements *v1, v2, ..., vn* with *r* rows and *c* columns, the *matrix* command is used in R with the following syntax:

M = matrix(c(v1, v2, v3, ... ,vn), nrow=r, ncol=c)

With this definition the elements of the matrix are placed consecutively by columns until completing the *n = rxc* elements that constitute it.

```
> M <- matrix(c(1,2,3,4,5,6),nrow = 2, ncol = 3)
> M
     [,1] [,2] [,3]
[1,]    1    3    5
[2,]    2    4    6
```

It is possible to dispense with the literals *nrow* and *ncol,* leaving the definition syntax of a dimension matrix (*r, c*) as follows:

M = matrix(c(v1, v2, v3, ,vn.......), r, c)

```
> M = matrix(c(1,2,3,4,5,6), 2, 3)
> M
     [,1] [,2] [,3]
[1,]    1    3    5
[2,]    2    4    6
```

If we want to create the array *by* rows we use the *byrow* command through the following syntax:

M = matrix(c(v1, v2, v3, ...,vn), byrow=TRUE)

```
> M = matrix(c(1,2,3,4,5,6),nrow = 2, ncol = 3, byrow=TRUE)
> M
     [,1] [,2] [,3]
[1,]    1    2    3
[2,]    4    5    6
```

If we want to add names to the column indexes: *Oranges, Bananas* and *Melon, and* in the rows: *Supermarket* and *Store*, we use the *dimnames* command through the following syntax:

```
> M = matrix(c(1,2,3,4,5,6),nrow = 2, ncol = 3, byrow=TRUE,
+ dimnames=list(c("Supermarket", "Store"), c("Oranges", +
"Bananas","Melon"))
```

```
> M
            Oranges Bananas Melon
Supermarket       1       2     3
Store             4       5     6
```

There are also specific commands for defining matrix variables, among which are the following:

rbind(x,y,z,...)	*It defines the matrix whose rows are the vectors x, y, z, ...*
	
	`> rbind(1:3,11:13,21:23)`
	`     [,1] [,2] [,3]`
	`[1,]    1    2    3`
	`[2,]   11   12   13`
	`[3,]   21   22   23`

cbind(x,y,z,...)	*It defines the matrix whose columns are the vectors x, y, z, ...*
	
	`> cbind(1:3,11:13,21:23)`
	`     [,1] [,2] [,3]`
	`[1,]    1   11   21`
	`[2,]    2   12   22`
	`[3,]    3   13   23`

array(x, dim=c(m,n))	*Defines the dimension matrix (m,n) with the elements of vector x by columns*
	
	`> M = array(c(1,2,3,4,5,6),dim=c(2,3)`
	`)`
	`> M`
	`     [,1] [,2] [,3]`
	`[1,]    1    3    5`
	`[2,]    2    4    6`

2.6.1 Matrix variable elements

R allows the selection of elements of matrix variables using the following commands:

A[m,n]	*Defines the element (m,n) of matrix A (row m and column n)*
	
	`> A=rbind(c(3,5,7,4), c(1,6,8,9), c(2,6,8,1))`
	`> A`
	`     [,1] [,2] [,3] [,4]`

	`[1,]      3      5      7      4`
	`[2,]      1      6      8      9`
	`[3,]      2      6      8      1`
	
	`> A[2,4]`
	`[1] 9`

A[a:b,c:d]	*Defines the submatrix of A formed by the rows between the a-th and b-th and by the columns between the c-th and d-th*
	
	`> A[1:2,2:3]`
	`      [,1] [,2]`
	`[1,]    5    7`
	`[2,]    6    8`

A[,c:d]	*Defines the sub-array of A formed by all the rows of A and by the columns between the c-th and d-th*
	
	`> A[,2:4]`
	`      [,1] [,2] [,3]`
	`[1,]    5    7    4`
	`[2,]    6    8    9`
	`[3,]    6    8    1`

A [a:b,]	*Defines the sub-array of A formed by all the columns of A and by the rows between the a-th and b-th*
	
	`> A[2:3,]`
	`      [,1] [,2] [,3] [,4]`
	`[1,]    1    6    8    9`
	`[2,]    2    6    8    1`

A[, c(a,b,c,...)]	*It defines the subatrix of A formed by the columns a- ith, b-ith, c- ith, etc. of matrix A.*
	
	`> A[,c(1,3,4)]`
	`      [,1] [,2] [,3]`
	`[1,]    3    7    4`
	`[2,]    1    8    9`
	`[3,]    2    8    1`

A[c(a,b,c,...),]	*It defines the submatrix of A formed by the rows a- ith, b-ith, c- ith, etc. of matrix A.*
	
	`> A[c(1,3),]`
	`      [,1] [,2] [,3] [,4]`
	`[1,]    3    5    7    4`
	`[2,]    2    6    8    1`

A[a,]	*Defines row a-number of matrix A*
	
	`> A[3,]`
	`[1] 2 6 8 1`

<table>
<tr><td>A[,b]</td><td>Defines the b-through column of Matrix A
<pre>
> A[,3]
[1] 7 8 8
</pre></td></tr>
<tr><td>A[,]</td><td>Equivalent to Matrix A
<pre>
> A[,]
     [,1] [,2] [,3] [,4]
[1,]   3   5   7   4
[2,]   1   6   8   9
[3,]   2   6   8   1
</pre></td></tr>
<tr><td>rbind(A,B,C,...)</td><td>It defines the matrix formed by the submatrices A, B, C, ... located in order one below the other in a consecutive way
<pre>
> M1=matrix(1:4,2,2)
> M1
     [,1] [,2]
[1,]   1   3
[2,]   2   4
> M2=matrix(5:8,2,2)
> M2
     [,1] [,2]
[1,]   5   7
[2,]   6   8
> M3=matrix(9:12,2,2)
> M3
     [,1] [,2]
[1,]   9   11
[2,]   10   12
> rbind(M1,M2,M3)
 [,1] [,2]
[1,]   1   3
[2,]   2   4
[3,]   5   7
[4,]   6   8
[5,]   9   11
[6,]   10   12
</pre></td></tr>
<tr><td>cbind(A,B,C,..)</td><td>It defines the matrix formed by the submatrices A, B, C, ... placed in order next to each other in a consecutive way
<pre>
> cbind(M1,M2,M3)
     [,1] [,2] [,3] [,4] [,5] [,6]
[1,]   1   3   5   7   9   11
[2,]   2   4   6   8   10   12
</pre></td></tr>
</table>

2.6.2 Specific matrix functions

MATLAB uses a group of predefined matrix functions that make working in the matrix field easier. The most important ones are the following:

<table>
<tr><td>diag(V)</td><td>Create a diagonal matrix with the vector V in the diagonal
<pre>
> diag(c(2,0,9,8,7))
     [,1] [,2] [,3] [,4] [,5]
[1,]   2   0   0   0   0
[2,]   0   0   0   0   0
[3,]   0   0   9   0   0
[4,]   0   0   0   8   0
[5,]   0   0   0   0   7
</pre></td></tr>
<tr><td>diag(A)</td><td>Extracts the diagonal from matrix A as a vector
<pre>
>A=rbind(c(1,3,5), c(2,0,8), c(-1,-3,2))
> A
     [,1] [,2] [,3]
[1,]   1   3   5
[2,]   2   0   8
[3,]   -1   -3   2
> diag(A)
[1] 1 0 2
</pre></td></tr>
<tr><td>diag(n)</td><td>Creates the Order Identity Matrix n
<pre>
> diag(4)
     [,1] [,2] [,3] [,4]
[1,]   1   0   0   0
[2,]   0   1   0   0
[3,]   0   0   1   0
[4,]   0   0   0   1
</pre></td></tr>
<tr><td>matrix(a,m,n)</td><td>Create the mxn order matrix with all elements of the matrix equal to the value a
<pre>
> matrix(1,4,5)
     [,1] [,2] [,3] [,4] [,5]
[1,]   1   1 1 1   1
[2,]   1   1   1   1   1
[3,]   1   1   1   1   1
[4,]   1   1   1   1   1
> matrix(8,4,5)
     [,1] [,2] [,3] [,4] [,5]
[1,]   8   8   8 8 8
[2,]   8   8   8 8 8
[3,]   8   8   8 8 8
[4,]   8   8   8 8 8
</pre></td></tr>
</table>

matrix(rnorm(a),m,n)	*Create the random matrix of normal elements with dimension (m,n) being a=mxn* > matrix(rnorm(9),3,3) [,1] [,2] [,3] [1,] 0.7753502 1.009282 -0.1153786 [2,] -0.9162865 1.280172 1.7335183 [3,] -0.2634779 -1.861913 0.1319272
dim(A)	*Returns the dimension of matrix A* > M=matrix(c(1,2,3,4,5,6),3,2) > M [,1] [,2] [1,] 1 4 [2,] 2 5 [3,] 3 6 > dim(M) [1] 3 2
colSums(A)	*Returns the sum of the columns in matrix A* > colSums(M) [1] 6 15
rowSums(A)	*Returns the sum of the rows in matrix A* > rowSums(M) [1] 5 7 9
t(A)	*Returns the Matrix Translated from Matrix A* > t(M) [,1] [,2] [,3] [1,] 1 2 3 [2,] 4 5 6
det(A)	*Returns the determinant of the square matrix A* >A=rbind(c(1,3,5), c(2,0,8), c(-1,-3,2)) > A [,1] [,2] [,3] [1,] 1 3 5 [2,] 2 0 8 [3,] -1 -3 2 > det(A) [1] -42
solve(A)	*Returns the inverse matrix of the square matrix A* > solve(A) [,1] [,2] [,3] [1,] -0.5714286 0.5000000 -0.57142857 [2,] 0.2857143 -0.1666667 -0.04761905 [3,] 0.1428571 0.0000000 0.14285714

solve(A,b)	*Resolves the Ax=b system*
	
	`> solve(A,c(1,2,3))`
	`[1] -1.2857143 -0.1904762 0.5714286`

eigen(A)	*Returns the auto values and auto vectors of matrix A*
	
	`> eigen(A)`
	`$values`
	`1] 2.17917+5.115792i 2.17917-5.115792i -1.35834+0.000000i`
	
	`$vectors`
	`[,1] [,2]`
	`[,3]`
	`1,] 0.5333555-0.2878809i 0.5333555+0.2878809i 0.8798028+0i`
	`2,] 0.6346542+0.0000000i 0.6346542+0.0000000i -0.4533254+0`
	`i`
	`3,] 0.0395385+0.4778151i 0.0395385-0.4778151i -0.1429795+0`
	`i`

kappa(A)	*Matrix Condition Number A*
	
	`> kappa(A)`
	`[1] 13.61015`

2.6.3 Matrix variable functions

The real single-variable and vector functions are expandable to matrix variable. To apply a simple variable or vector function to a matrix variable, just apply the function to each element of the matrix. Below are examples of trigonometric and inverse trigonometric functions.

Function	*Reverse*

sin(A) *Sine*
 	
`>A=rbind(c(1,3,5),c(2,0,8),c(-1,-3,2))`
`> A`
`[,1] [,2] [,3]`
`[1,] 1 3 5`
`[2,] 2 0 8`
`[3,] -1 -3 2`
 	
`> without(A)`
`[,1] [,2] [,3]`
`[1,] 0.8414710 0.14112 -0.9589243`
`[2,] 0.9092974 0.00000 0.9893582`
`[3,] -0.8414710 -0.14112 0.9092974`
	asin(A) *Sine arch*
	
	`> asin(A)`
	`[,1] [,2] [,3]`
	`[1,] 1.570796 NaN NaN`
	`[2,] NaN 0 NaN`
	`[3,] -1.570796 NaN NaN`

cos(A) *Cosine*
`> cos(A)`
`[,1] [,2] [,3]`
`[1,] 0.5403023 -0.9899925 0.2836622`
`[2,] -0.4161468 1.0000000 -0.1455000`
`[3,] 0.5403023 -0.9899925 -0.4161468`
	acos(A) *Cosine arc*
	`> acos(A)`
	`[,1] [,2] [,3]`
	`[1,] 0.000000 NaN NaN`
	`[2,] NaN 1.570796 NaN`
	`[3,] 3.141593 NaN NaN`

tan(A) *Tangent*

```
> tan(A)
          [,1]       [,2]      [,3]
[1,]  1.557408 -0.1425465 -3.380515
[2,] -2.185040  0.0000000 -6.799711
[3,] -1.557408  0.1425465 -2.185040
```

atan(A) and atan2(A)

```
> atan(A)
           [,1]      [,2]     [,3]
[1,]  0.7853982  1.249046 1.373401
[2,]  1.1071487  0.000000 1.446441
[3,] -0.7853982 -1.249046 1.107149
```

sinh(A) *Hyperbolic sine*

```
> sinh(A)
          [,1]      [,2]       [,3]
[1,]  1.175201  10.01787   74.20321
[2,]  3.626860   0.00000 1490.47883
[3,] -1.175201 -10.01787    3.62686
```

asinh(A) *Hyperbolic sine arc*

```
> asinh(A)
           [,1]      [,2]     [,3]
[1,]  0.8813736  1.818446 2.312438
[2,]  1.4436355  0.000000 2.776472
[3,] -0.8813736 -1.818446 1.443635
```

cosh(A) *Hyperbolic cosine*

```
> cosh(A)
         [,1]     [,2]        [,3]
[1,] 1.543081 10.06766   74.209949
[2,] 3.762196  1.00000 1490.479161
[3,] 1.543081 10.06766    3.762196
```

acosh(A) *Hyperbolic cosine arc*

```
> acosh(A)
         [,1]     [,2]     [,3]
[1,] 0.000000 1.762747 2.292432
[2,] 1.316958      NaN 2.768659
[3,]      NaN      NaN 1.316958
```

tanh(A) *Hyperbolic Tangent*

```
> tanh(A)
           [,1]       [,2]      [,3]
[1,]  0.7615942  0.9950548 0.9999092
[2,]  0.9640276  0.0000000 0.9999998
[3,] -0.7615942 -0.9950548 0.9640276
```

atanh(A) *Hyperbolic arc tangent*

```
> atanh(A)
     [,1] [,2] [,3]
1,]  Inf   NaN  NaN
2,]  NaN     0  NaN
3,] -Inf   NaN  NaN
```

Below are examples of exponential, logarithmic, and other vector-variable functions:

Function	***Meaning***
exp(A)	*Exponential function based on e of the elements of matrix A*

```
> A=rbind(c(1,3,5), c(2,0,8), c(-1,-3,2
))
> A
     [,1] [,2] [,3]
[1,]    1    3    5
[2,]    2    0    8
[3,]   -1   -3    2
> exp(A)
          [,1]        [,2]        [,3]
[1,] 2.7182818 20.08553692  148.413159
[2,] 7.3890561  1.00000000 2980.957987
[3,] 0.3678794  0.04978707    7.389056
```

<table>
<tr><td>log(A)</td><td>Logarithm function on the basis of e of the elements of matrix A
<pre>
> log(A)
               [,1]   [,2]   [,3]
[1,] 0.0000000 1.098612 1.6094379
[2,] 0.6931472   -Inf 2.0794415
[3,]   NaN   NaN 0.6931472
</pre></td></tr>
<tr><td>log10(A)</td><td>Logarithm function on the basis of 10 of the elements of matrix A
<pre>
> log10(A)
        [,1]   [,2]   [,3]
[1,] 0.00000 0.4771213 0.69897
[2,] 0.30103   -Inf 0.90309
[3,]   NaN   NaN 0.30103
</pre></td></tr>
<tr><td>log2(A)</td><td>Logarithm function based on 2 of the elements of the A matrix
<pre>
> log2(A)
      [,1]   [,2]   [,3]
[1,]   0 1 .584963 2.321928
[2,]   1   -Inf 3.000000
[3,]   NaN   NaN 1.000000
</pre></td></tr>
<tr><td>sqrt(A)</td><td>Square root function of the elements of the matrix A
<pre>
sqrt(A)
             [,1]   [,2]   [,3]
[1,] 1.000000 1.732051 2.236068
[2,] 1.414214 0.000000 2.828427
[3,]   NaN   NaN 1.414214
</pre></td></tr>
<tr><td>floor(A)</td><td>The largest integer less than or equal to each element of matrix A
<pre>
> floor(sqrt(A))
      [,1] [,2] [,3]
[1,]   1   1   2
[2,]   1   0   2
[3,]   NaN NaN   1
</pre></td></tr>
<tr><td>ceiling(A)</td><td>The smallest integer greater than or equal to each element of matrix A
<pre>
> ceiling(sqrt(A))
      [,1] [,2] [,3]
[1,]   1   2   3
[2,]   2   0   3
[3,]   NaN NaN   2
</pre></td></tr>
</table>

round(A)	*The integer closest to each element of V*
	
	`> round(sqrt(A))`
	`      [,1] [,2] [,3]`
	`[1,]     1    2    2`
	`[2,]     1    0    3`
	`[3,]   NaN  NaN    1`

trunc(A)	*Removes the decimal part of each V*
	
	`> trunc(sqrt(A))`
	`      [,1] [,2] [,3]`
	`[1,]     1    1    2`
	`[2,]     1    0    2`
	`[3,]   NaN  NaN    1`

abs(A)	*Absolute value of each V element*
	
	`> A=rbind(c(1,3,5), c(2,0,8), c(-1,-3,2`
	`))`
	`> A`
	`      [,1] [,2] [,3]`
	`[1,]     1    3    5`
	`[2,]     2    0    8`
	`[3,]    -1   -3    2`
	
	`> abs(A)`
	`      [,1] [,2] [,3]`
	`[1,]     1    3    5`
	`[2,]     2    0    8`
	`[3,]     1    3    2`

sign(A)	*Sign of each V element (1 if $V_i>0$ and -1 if $V_i<0$)*
	
	`> sign(A)`
	`      [,1] [,2] [,3]`
	`[1,]     1    1 1`
	`[2,]     1    0    1`
	`[3,]    -1   -1    1`

2.7 OPERATORS

R is a language that incorporates arithmetic, logical and relational operators just like any other language. The following are the types of operators cited in the scope of the R language.

2.7.1 Arithmetic operators

R incorporates the usual arithmetic operators (addition, subtraction, multiplication and division) to work with numbers. But at the same time it extends the sense of these operators to work with scalars, vectors and matrices, as shown in the following table:

Operator	***Function that it carries out***

+	*Sum of scalars, vectors or matrices*
	`> A=c(1,3,-2,6);B=c(4,-5,8,2); c=3;`
	`> V1=A+c`
	`> V1`
	`[1] 4 6 1 9`
	`> V2=A+B`
	`> V2`
	`[1] 5 -2 6  8`

-	*Subtraction of scalars, vectors or matrices*
	`> V3=A-c`
	`> V3`
	`[1] -2 0 -5 3`
	`> V4=A-B`
	`> V4`
	`[1] -3   8 -10    4`

*	*Product of scalars or matrices (element by element) or of scalars by vectors or matrices*
	`> V5=A*c`
	`> V5`
	`[1] 3  9 -6 18`

^ or **	*Scaling power of a scalar, a vector or an array (in the case of the vector or array, each element is raised when scaling)*
	`> V5^2`
	`[1]    9 81 36 324`
	`> V5**2`
	`[1]    9 81 36 324`
	`> c^3`
	`[1] 27`
	`> c**3`
	`[1] 27`

%*%	*Matrix product (matrix product)*
	`> A=matrix(rnorm(4),2,2)`
	`> A`
	`           [,1]       [,2]`
	`[1,] 1.082627 -1.4716030`
	`[2,] -2.193202 0.7243937`

```
> B=matrix(rnorm(4),2,2)
> B
            [,1]       [,2]
[1,] 0.08075123 -1.689854
[2,] -0.06904861 1.087078
> A%*%B
           [,1]       [,2]
[1,] 0.1890356 -3.429229
[2,] -0.2271221 4.493664
```

2.7.2 Logical operators

R also incorporates the usual logical operators using the most common notation for them. Usually the results of the logical operators are usually 1 if true and 0 if false. The following table shows these operators.

<table>
<tr><td>!=</td><td>Logical denial (NOT) or complementary to A

<code>> 2!=1</code>
<code>1] TRUE</code></td></tr>
<tr><td>A & B</td><td>Logical Conjunction (AND) or intersection of A and B

<code>> (2>3) & (5>1)</code>
<code>1] FALSE</code></td></tr>
<tr><td>A | B</td><td>Logical disjunction (OR) or union of A and B

<code>> (2>3)||(5>1)</code>
<code>1] TRUE</code></td></tr>
<tr><td>xor(A,B)</td><td>Exclusive OR (XOR) or symmetrical difference of A and B (TRUE voucher if A or B, but not both, are TRU)

<code>> xor((2>3),(5>1))</code>
<code>1] TRUE</code></td></tr>
<tr><td>any(A,B,C,...)</td><td>It is TRUE if at least one component A, B, C,... is TRUE

<code>> any((2>3),(5>1),(4>5))</code>
<code>1] TRUE</code></td></tr>
<tr><td>all(A,B,C,...)</td><td>It is TRUE if all the components A, B, C,... are TRUE

<code>> all((2>3),(5>1),(4>5))</code>
<code>1] FALSE</code></td></tr>
</table>

2.7.3 Relational operators

Likewise, R deals with relational operations that execute element to element comparisons between two arrays and return an array of the same size whose elements are zeros if the corresponding relationship is true, or ones if the corresponding relationship is false. Relational operators also allow to compare scalars with vectors or arrays, in which case the scalar is compared with all the elements of the array. The following table shows the relational operators of R.

<	*Minor (for complexes it only affects real parts)*
	`> 3<5`
	`1] TRUE`

<=	*Minor or equal (for complexes it only affects real parts)*
	`> 4>=6`
	`1] FALSE`

>	*Major (for complexes it only affects real parts)*
	`> X=matrix(3,4,2)`
	`> X`
	`      [,1] [,2]`
	`[1,]    3    3`
	`[2,]    3    3`
	`[3,]    3    3`
	`[4,]    3    3`
	`> Y=matrix(c(1,2,3,4,5,6,7,8),4,2)`
	`> Y`
	`      [,1] [,2]`
	`[1,]    1    5`
	`[2,]    2    6`
	`[3,]    3    7`
	`[4,]    4    8`
	`> X>Y`
	`       [,1] [,2]`
	`1,] TRUE FALSE`
	`2,] TRUE FALSE`
	`3,] FALSE FALSE`
	`4,] FALSE FALSE`
	
	`> Y>3`
	`       [,1] [,2]`
	`1,] FALSE TRUE`
	`2,] FALSE TRUE`
	`3,] FALSE TRUE`
	`4,] TRUE TRUE`

>=	*Greater or equal (for complexes it only affects real parts)*
	`> X>=Y`
	`       [,1] [,2]`
	`1,] TRUE FALSE`
	`2,] TRUE FALSE`
	`3,] TRUE FALSE`
	`4,] FALSE FALSE`

x == y	*Equality (affects complex numbers)*
	
	`> X==Y`
	`       [,1] [,2]`
	`1,] FALSE FALSE`
	`2,] FALSE FALSE`
	`3,] TRUE FALSE`
	`4,] FALSE FALSE`

x ! = y	*Inequality (affects complex numbers)*
	
	`> X!=Y`
	`       [,1] [,2]`
	`1,] TRUE TRUE`
	`2,] TRUE TRUE`
	`3,] FALSE TRUE`
	`4,] TRUE TRUE`

2.8 INFORMATION FUNCTIONS

R has several functions whose purpose is to inform about the nature of your argument. For example, the function *is.numeric(x)* will return TRUE if *x is* a numerical value, and will return FALSE otherwise.

```
> v=4
> v
[1] 4
> is.numeric(v)
1] TRUE
```

The function *is. matrix(x)* will return TRUE if *x is* an array and return FALSE otherwise.

```
> M = matrix(c(1,2,3,4,5,6), 2, 3)
> M
      [,1] [,2] [,3]
[1,]     1    3    5
[2,]     2    4    6
> is.matrix(M)
1] TRUE
```

To obtain the list of information functions contained in R, simply type in the *method(s).*

```
> methods(is)
 1] is.array          is.atomic        is.call
 4] is.character      is.complex       is.data.frame
 7] is.double         is. element      is.empty.model
10] is.environment    is.expression    is.factor
```

```
13] is.finite             is.function              is.infinite
16] is.integer            is.language              is.leaf
19] is.list               is.loaded                is.logical
22] is.matrix             is.mts                   is.na
25] is.na.data.frame      is.na.numeric_version    is.na.POSIXlt
28] is.na<-               is.na<-.default          is.na<-.factor
31] is.name               is.nan                   is.null
34] is.numeric            is.numeric.Date          is.numeric.difftime
37] is.numeric.POSIXt     is.numeric_version       is.object
40] is.ordered            is.package_version       is.pairlist
43] is.primitive          is.qr                    is.R
46] is.raster             is.raw                   is.recursive
49] is.relistable         is.single                is.stepfun
52] is.symbol             is.table                 is.ts
55] is.tskernel           is.unsorted              is.vector
```

2.9 CONVERSION FUNCTIONS

The conversion functions transform a variable of a certain type to another specified type. For example, if we want to transform the numerical variable *v=4* into a *character variable* we will use the function *as.character(v).*

```
> v=4
> v
[1] 4
> as.character(v)
[1] "4"
```

To obtain the list of conversion functions contained in R, just type in *methods(as).*

```
> methods(as)
 1] as.array                         as.array.default
 3] as.call                          as.character
 5] as.character.condition          .as.character.Date
 7] as.character.default             as.character.error
 9] as.character.factor              as.character.hexmode
 11] as.character.numeric_version    as.character.octmode
 13] as.character.POSIXt             as.character.srcref
 15] as.complex                      as.data.frame
 17] as.data.frame.array             as.data.frame.AsIs
 19] as.data.frame.character         as.data.frame.complex
 21] as.data.frame.data.frame        as.data.frame.Date
 23] as.data.frame.default           as.data.frame.difftime
 25] as.data.frame.factor            as.data.frame.integer
 27] as.data.frame.list              as.data.frame.logical
 29] as.data.frame.matrix            as.data.frame.model.matrix
 31] as.data.frame.numeric           as.data.frame.numeric_version
 33] as.data.frame.ordered           as.data.frame.POSIXct
```

```
 35] as.data.frame.POSIXlt          as.data.frame.raw
 37] as.data.frame.table            as.data.frame.ts
 39] as.data.frame.vector          .as.Date
 41] as.Date.character              as.Date.date
 43] as.Date.dates                  as.Date.default
 45] as.Date.factor                .as.Date.numeric
 47] as.Date.POSIXct                as.Date.POSIXlt
 49] as.dendrogram                  as.difftime
 51] as.dist                        as.double
 53] as.double.difftime             as.double.POSIXlt
 55] as.environment                 as.expression
 57] as.expression.default          as.factor
 59] as.formula                     as.function
 61] as.function.default            as.graphicsAnnot
 63] as.hclust                      as.hexmode
 65] as.integer                     as.list
 67] as.list.data.frame             as.list.Date
 69] as.list.default                as.list.environment
 71] as.list.factor                 as.list.function
 73] as.list.numeric_version        as.list.POSIXct
 75] as.logical                     as.logical.factor
 77] as.matrix                      as.matrix.data.frame
 79] as.matrix.default              as.matrix.noquote
 81] as.matrix.POSIXlt              as.name
 83] as.null                        as.null.default
 85] as.numeric                     as.numeric_version
 87] as.octmode                     as.ordered
 89] as.package_version             as.pairlist
 91] as.person                      as.personList
 93] as.POSIXct                     as.POSIXct.date
 95] as.POSIXct.Date                as.POSIXct.dates
 97] as.POSIXct.default             as.POSIXct.numeric
 99] as.POSIXct.POSIXlt             as.POSIXlt
101] as.POSIXlt.character           as.POSIXlt.date
103] as.POSIXlt.Date                as.POSIXlt.dates
105] as.POSIXlt.default             as.POSIXlt.factor
107] as.POSIXlt.numeric             as.POSIXlt.POSIXct
109] as.qr                          as.raster
111] as.raw                         as.relistable
113] as.roman                       as.single
115] as.single.default              as.stepfun
117] as.symbol                      as.table
119] as.table.default               as.ts
121] as.vector                      as.vector.factor
```

2.10 VARIABLES AND CHAIN FUNCTIONS

A character variable (string) is simply a string of characters enclosed in single quotes that R treats as a vector. The general syntax for the declaration of character variables in R is as follows:

c = "string of characters”

```
> c="string of characters
> c
1] "string of characters
```

Among the R functions that handle character variables we have the following:

nchar(string_character)	*Returns the number of characters in the string*
	
	`> c`
	`1] "chain of characters"`
	`> c="string of characters″`
	`> nchar(c)`
	`[1] 20`

substring(x, n)	*Returns the substring of string x that starts at place character n and ends with the string*
	
	`> substring(c,10)`
	`1] " characters`

substring(x, n, m)	*Returns the substring of string x that starts at place character n and ends at place character m*
	
	`> substring(c,5,15)`
	`1] "na de carac"`

strsplit(x, "string")	*Divide string x into a list of separate sub-strings delimited by the string "string*
	
	`> strsplit(c, "e")`
	`[[1]]`
	`1] "cad" "na d" " charact"`
	`"r" "s"`

grep("pattern", "chain")	*Checks if the specified pattern is contained in the string, in which case it returns 1*
	
	`> grep("c",c)`
	`[1] 1`
	
	`> grep("b",c)`
	`integer(0)`

<table>
<tr><td>regexpr("pattern", "string")
ó
regexec("pattern", "string")</td><td>Returns the position in the chain of the first occurrence of the given pattern within the chain

<code>> regexpr("a", c)
[1] 2
attr(,"match.length")
[1] 1
attr(,"useBytes")
1] TRUE</code>

<code>> regexpr("d",c)
[1] 3
attr(,"match.length")
[1] 1
attr(,"useBytes")
1] TRUE</code>

<code>> regexec("de",c)
[[1]]
[1] 3
attr(,"match.length")
[1] 2</code>

<code>> regexec("ca",c)
[[1]]
[1] 1
attr(,"match.length")
[1] 2</code></td></tr>
<tr><td>gregexpr("pattern", "string")</td><td>Returns all positions in the chain of occurrences of the given pattern within the chain

<code>> gregexpr("a", c)
[[1]]
[1] 2 6 12 14
attr(,"match.length")
[1] 1 1 1 1
attr(,"useBytes")
1] TRUE</code></td></tr>
<tr><td>sub("pattern", "replacement value", "string")</td><td>Replaces the specified pattern in the string with the replacement value for the first occurrence

<code>> sub("a","x",c)
1] "cxdensity of characters"</code></td></tr>
</table>

gsub("pattern", "replacement value", "string")	*Replaces the specified pattern in the string with the replacement value for all occurrences*
	
	`> gsub("a","x",c)`
	`1] "cxdenx de cxrxcteres"`

tolower ("chain")	*Converts all the characters in the chain to lower case*
	
	`> tolower("CHAIN")`
	`1] "chain"`

toupper ("chain")	*Converts all the characters in the chain to capital letters*
	
	`> toupper("chain")`
	`1] "CHAIN"`

match("chain",table)	*Returns a vector with the position of the first exact match of the whole string in the table*
	
	`> match("mean", c("median", "mean", "mode", "mean"))`
	`[1] 2`

pmatch("chain",table)	*Returns a vector with the position of the first match of part of the string in the table*
	
	`> pmatch("med", c("mean", "median", "mode"))`
	`[1] 2`

paste("string", array, sep= "separation characters")	*Places the string as elements of the given array separated by the specified separation character*
	
	`> paste("A", 1:6, sep = "")`
	`1] "A1" "A2" "A3" "A4" "A5" "A6"`

2.11 SELECTION AND HANDLING FUNCTIONS

Below is a group of functions used to select and manipulate the elements of a vector or array.

which.max(v)	*Returns the index of the largest V element*
	
	`> V=c(10,25,3,8,49,12,3,NA)`
	`> V`
	`1] 10 25 3 8 49 12 3 NA`
	`> which.max(V)`
	`[1] 5`

which.min(v)	*Returns the index of the lowest element of V*
	
	`> which.min(V)`
	`[1] 3`

which(v==a)	*Returns the indexes of the v-elements that match a*
	
	`> which(V==3)`
	`[1] 3 7`

rev(v)	*Reverses the order of the elements of V*
	
	`> rev(V)`
	`1] NA 3 12 49 8 3 25 10`

sort(v)	*Sort the V elements in increasing order*
	
	`> sort(V)`
	`[1] 3  3 8 10 12 25 49`

rev(sort(v))	*Sort the V elements in decreasing order*
	
	`> rev(sort(V))`
	`[1] 49 25 12 10 8  3 3`

unique(v)	*Returns the vector without repeated elements*
	
	`> unique(V)`
	`1] 10 25 3 8 49 12 NA`

subset(v, condition)	*Returns v-elements that meet the condition*
	
	`> subset(V, V<10)`
	`[1] 3 8 3`

sample(v,n)	*Returns a simple random sample without replacing n elements of the vector v*
	`> sample(V,4)`
	`[1] 49 3  8 10`

sample(v,n, replace=TRUE)	*Returns a simple random sample with replacement of n elements of the vector v*
	`> sample(V,4,replace=TRUE)`
	`[1] 25 49 8  3`

table(v)	*Returns a table with the frequencies of the different v*
	`> table(V)`
	`V`
	` 3 8 10 12 25 49`
	` 2 1  1 1  1 1`

na.omit(v)	*Suppresses observations with missing values of v*
	`> na.omit(V)`
	`[1] 10 25 3  8 49 12 3`
	`attr(,"na.action")`
	`[1] 8`
	`attr(,"class")`
	`1] "omit"`

na.fail(v)	*Returns an error message if v contains any missing elements*
	`> na.fail(V)`
	`Error in na.fail.default(V) : missi`
	`ng values in object`

The above selection and manipulation functions can also be applied to matrices and arrays. Examples are presented below:

```
> A=matrix(c(10,25,3,40,5,6,7,8,9),3,3)
> A
     [,1] [,2] [,3]
[1,]   10   40    7
[2,]   25    5    8
[3,]    3    6    9
```

```
> which.max(A)
[1] 4

> which.min(A)
[1] 3

> rev(A)
[1] 9  8 7  6 5 40 3 25 10

> sort(A)
[1] 3  5 6  7 8  9 10 25 40

> rev(sort(A))
[1] 40 25 10 9  8 7  6 5  3

> which(A==5)
[1] 5

> which(A==25)
[1] 2

> na.omit(A)
      [,1] [,2] [,3]
[1,]    10   40    7
[2,]    25    5    8
[3,]     3    6    9

> na.fail(A)
      [,1] [,2] [,3]
[1,]    10   40    7
[2,]    25    5    8
[3,]     3    6    9

> table(A)
A
 3 5  6 7  8 9 10 25 40
 1 1  1 1  1 1  1 1  1

> subset(A<10)
        [,1] [,2] [,3]
1,] FALSE FALSE TRUE
2,] FALSE TRUE
3,] TRUE TRUE

> sample(A,4)
[1] 5 8 9 7
```

Calculate the default value of R for the factorial of 23. Calculate it also with 15, with 20 and with 22 significant figures.

```
> factorial(23)
[1] 2.585202e+22

> options(digits=15)
> factorial(23)
[1] 2.5852016738885e+22

> options(digits=20)
> factorial(23)
[1] 25852016738884978212844

> options(digits=22)
> factorial(23)
[1] 25852016738884978212844
```

*Given the vector variables a=[π ,2π ,3π ,4π ,5 π] and b=[e,2e,3e,4e,5e] calculate c=Sen(a)+b, d=Cosh(a), e=Ln(b), f=c*d, g=c/d, h=d2. The products, kitchens and powers are considered element by element*

```
> a=c(pi,2*pi,3*pi,4*pi,5*pi)
> b=c(exp(1),2*exp(1),3*exp(1),4*exp(1),5*exp(1))
> a
[1] 3.141593 6.283185 9.424778 12.566371 15.707963
> b
[1] 2.718282 5.436564 8.154845 10.873127 13.591409


> c=sin(a)+b
> c
[1] 2.718282 5.436564 8.154845 10.873127 13.591409

> d=cosh(a)
> d
1] 1.159195e+01 2.677468e+02 6.195824e+03 1.433757e+05 3.3178
12e+06
> e=log(b)
> e
[1] 1.000000 1.693147 2.098612 2.386294 2.609438

> f=c*d
> f
1] 3.151020e+01 1.455622e+03 5.052599e+04 1.558942e+06 4.5093
74e+07
```

```
> g=c/d
> g
1] 2.344973e-01 2.030487e-02 1.316184e-03 7.583663e-05 4.0964
98e-06

> h=d^2
> h
1] 1.343734e+02 7.168833e+04 3.838823e+07 2.055658e+10 1.1007
88e+13
```

Exercise 3. Given the vector of the first 10 natural numbers, find:

Its sixth element
Its elements located between the fourth and the seventh both inclusive
Its elements located between the second and the ninth both inclusive of three
The elements of the previous section, but from greater to lesser

```
> x=(1:10)
> x
 [1] 1  2 3  4 5  6 7  8 9 10

> x[6]
[1] 6
```

We have obtained the sixth element of vector x.

```
> x[4:7]
[1] 4 5 6 7
```

We have obtained the elements of vector x located between the fourth and seventh, both inclusive.

```
> seq(2,9, by=3)
[1] 2 5 8
```

We have obtained the elements of vector x located between the second and the ninth, both included, but separated by three in three units.

```
> seq(9,2, by=-3)
[1] 9 6 3
```

We have obtained the elements of vector *x* located between the ninth and the second, both included, but separated by three in three units and starting from the ninth.

Exercise 4. Given the 2x3 matrix whose rows are the first six consecutive odd numbers, it is requested:

Cancel its element (2,3), transpose it and attach the identity matrix of order 3 to its right.

If we call C to the previously obtained matrix, build a D matrix extracting the odd columns of the C matrix, an E matrix formed by the intersection of the first two rows of C and its third and fifth columns, and an F matrix formed by the intersection of the first two rows and the last three columns of the C matrix.

Construct the diagonal G matrix such that the elements of its main diagonal are the same as those of the main diagonal of D.

We build the H matrix, formed by the intersection of the first and third rows of C and its second, third and fifth columns.

We first consider the *2x3* matrix whose rows are the first 6 consecutive odd numbers:

```
> A=matrix(c(1,3,5,7,9,11), 2,3,byrow=TRUE)
> A
     [,1] [,2] [,3]
[1,]    1    3    5
[2,]    7    9   11
```

Now we are going to cancel the elemennto *(2,3), that* is, its last element:

```
> A[2,3]=0
> A
     [,1] [,2] [,3]
[1,]    1    3    5
[2,]    7    9    0
```

Next, we consider the matrix *B* transposed from *A*:

```
> B=t(A)
> B
     [,1] [,2]
[1,]    1    7
[2,]    3    9
[3,]    5    0
```

Now we build an array *C,* formed by array *B* and the order identity array 3 attached to its right:

```
> C=cbind(B,diag(3))
> C
```

```
     [,1] [,2] [,3] [,4] [,5]
[1,]    1    7    1    0    0
[2,]    3    9    0    1    0
[3,]    5    0    0    0    1
```

We are going to construct a *D array by* extracting the odd columns of the *C array*, an *E array* formed by the intersection of the first two rows of *C and* its third and fifth columns, and an *F array* formed by the intersection of the first two rows and the last three columns of the *C array*:

```
> D=C[,c(1,3,5)]
> D
     [,1] [,2] [,3]
[1,]    1    1    0
[2,]    3    0    0
[3,]    5    0    1

> E=C[c(1,2),c(3,5)]
> E
     [,1] [,2]
[1,]    1    0
[2,]    0    0

> F=C[c(1,2),c(3,4,5)]
> F
     [,1] [,2] [,3]
[1,]    1    0    0
[2,]    0    1    0
```

Now we build the diagonal *G* matrix such that the elements of its main diagonal are the same as those of the main diagonal of *D*:

```
> G=diag(diag(D))
> G
     [,1] [,2] [,3]
[1,]    1    0    0
[2,]    0    0    0
[3,]    0    0    1
```

Next, we build the *H* matrix, formed by the intersection of the first and third rows of *C and* its second, third and fifth columns:

```
> H=C[c(1,3),c(2,3,5)]
> H
     [,1] [,2] [,3]
[1,]    7    1    0
[2,]    0    0    1
```

Exercise 5. To construct a matrix I formed by the matrix identity of order 5x4 and the matrices null and unit of the same order attached to its right. Then, extract the first row of I and, finally, form the matrix J with the odd rows and the even columns of Y and calculate its order (size).

Additionally, build a 3x4 order random K matrix and invert first the order of its rows, then the order of its columns and then the order of its rows and columns at the same time.

Now we build an *I* array formed by the *5x4* order identity array and the null and unit arrays of the same order attached to its right. Then, we extract the first row of *I and*, finally, we form the *J array* with the odd rows and the even columns of *Y and* we calculate its order (size).

```
> I=cbind(diag(5)[,c(1,2,3,4)],matrix(0,5,4),matrix(1,5,4))
> I
     [,1] [,2] [,3] [,4] [,5] [,6] [,7] [,8] [,9] [,10] [,11] [,12]
[1,]    1    0    0    0    0    0    0    0    1     1     1     1
[2,]    0    1    0    0    0    0    0    0    1     1     1     1
[3,]    0    0    1    0    0    0    0    0    1     1     1     1
[4,]    0    0    0    1    0    0    0    0    1     1     1     1
[5,]    0    0    0    0    0    0    0    0    1     1     1     1

> I[1,]
 [1] 1 0 0 0 0 0 0 0 1 1 1 1

> J=I[c(1,3,5),c(2,4,6,8,10,12)]
> J
     [,1] [,2] [,3] [,4] [,5] [,6]
[1,]    0    0    0    0    1    1
[2,]    0    0    0    0    1    1
[3,]    0    0    0    0    1    1

> dim(J)
[1] 3 6
```

Next, we build a *3x4* order random *K* matrix and we reverse first the order of its rows, then the order of its columns and then the order of its rows and columns at the same time. Finally, we find the matrix *L of* order *4x3* whose columns result from taking the elements of the columns of *K* consecutively.

```
> K=matrix(rnorm(12),3,4)
> K
          [,1]        [,2]       [,3]       [,4]
[1,] 0.0367322 -0.07320238 -0.5295449 -0.9802087
[2,] 1.0106856 -1.67161887  1.0675619 -2.0160255
[3,] 0.1834108 -0.15990453  0.6908918 -1.4977978
```

```
> K[c(3,2,1),]
          [,1]        [,2]       [,3]       [,4]
[1,] 0.1834108 -0.15990453  0.6908918 -1.4977978
[2,] 1.0106856 -1.67161887  1.0675619 -2.0160255
[3,] 0.0367322 -0.07320238 -0.5295449 -0.9802087

> K[,c(4,3,2,1)]
           [,1]       [,2]        [,3]      [,4]
[1,] -0.9802087 -0.5295449 -0.07320238 0.0367322
[2,] -2.0160255  1.0675619 -1.67161887 1.0106856
[3,] -1.4977978  0.6908918 -0.15990453 0.1834108

> K[c(3,2,1),c(4,3,2,1)]
           [,1]       [,2]        [,3]      [,4]
[1,] -1.4977978  0.6908918 -0.15990453 0.1834108
[2,] -2.0160255  1.0675619 -1.67161887 1.0106856
[3,] -0.9802087 -0.5295449 -0.07320238 0.0367322
```

Exercice 6. Given the square matrix of order 3, whose rows are the first 9 natural numbers, obtain their inverse, their transposed and their diagonal. Obtain the sum of the elements of the first row and the sum of the elements of the diagonal. Extract the submatrix whose diagonal is the elements a11 and a22 and also extract the submatrix whose elements of the diagonal are a11 and a33. Finally, find the autovalues and autovectors of the initial matrix.

```
> M=matrix(c(1,2,3,4,5,6,7,8,9), 3,3,byrow=TRUE)
> M
     [,1] [,2] [,3]
[1,]    1    2    3
[2,]    4    5    6
[3,]    7    8    9
> A=solve(M)
Error in solve.default(M) :
 system is computationally singular: reciprocal condition num
ber = 1.54198e-18
```

The matrix has no inverse because its determinant is zero

```
> round(det(M))
[1] 0
```

The following sections of the exercise are resolved below.

```
> B=t(M)

> B
```

```
      [,1] [,2] [,3]
[1,]     1    4    7
[2,]     2    5    8
[3,]     3    6    9

> V=diag(M)
> V
[1] 1 5 9

> s=M[1,1]+M[1,2]+M[1,3]
> s
[1] 6
> sd=M[1,1]+M[2,2]+M[3,3]

> sd
[1] 15
> SM=M[c(1,2),c(1,2)]
> SM
      [,1] [,2]
[1,]     1    2
[2,]     4    5

> SM1=M[c(1,3),c(1,3)]
> SM1
      [,1] [,2]
[1,]     1    3
[2,]     7    9

> eigen(M)
$values
1] 1.611684e+01 -1.116844e+00 -1.303678e-15

$vectors
           [,1]        [,2]       [,3]
[1,] -0.2319707 -0.78583024  0.4082483
[2,] -0.5253221 -0.08675134 -0.8164966
[3,] -0.8186735  0.61232756  0.4082483
```

Exercice 7. Given the square matrix of order 3, whose rows are the first 9 natural numbers (without the zero), identify their values lower than 5.

If we now consider the vector whose elements are the first 9 natural numbers (without the zero) and the vector with the same elements from highest to lowest, identify the values resulting from subtracting from the elements of the second vector the quantity 1 if the corresponding element of the first vector is greater than 2, or the quantity 0 if it is less than or equal to 2.

```
> X=5*diag(3)
> X
     [,1] [,2] [,3]
[1,]    5    0    0
[2,]    0    5    0
[3,]    0    0    5

> X>=matrix(c(1,2,3,4,5,6,7,8,9),3,3,byrow=TRUE)
      [,1]  [,2]  [,3]
[1,]  TRUE FALSE FALSE
[2,] FALSE  TRUE FALSE
[3,] FALSE FALSE FALSE
```

The elements of matrix *X* that are greater or equal to those of the matrix [1 2 3; 4 5 6; 7 8 9] correspond to TRUE in the response matrix. The rest of the elements correspond to FALSE (the result of the operation would be the same if we compare the scalar 5 with the matrix [1 2 3; 4 5 6; 7 8 9].

```
> X=5
> X>=matrix(c(1,2,3,4,5,6,7,8,9),3,3,byrow=TRUE)
      [,1]  [,2]  [,3]
[1,]  TRUE  TRUE  TRUE
[2,]  TRUE  TRUE FALSE
[3,] FALSE FALSE FALSE
```

Next, we solve the second part of the exercise:

```
> A=1:9
> A
[1] 1 2 3 4 5 6 7 8 9

> B=10-A
> B
[1] 9 8 7 6 5 4 3 2 1

> Y=A>4
> Y
1] FALSE FALSE FALSE TRUE TRUE TRUE
> Z=B-(A>2)
> Z
[1] 9 8 6 5 4 3 2 1 0
```

The values of *Y* equal to TRUE correspond to the elements of A greater than 4. The values of *Z* result from subtracting from the corresponding elements of *B the* quantity 1 if the corresponding element of *A is* greater than 2, or the quantity 0 if the corresponding element of *A is* less than or equal to 2.

Exercice 8. If we now consider the vector A whose elements are the first 9 natural numbers (without the zero), identify its values greater than 2 and less than 6.

If we call P to the vector of the same dimension as A that associates a 1 to the results identified in the previous section and a zero to the rest, identify when A or P, but not both, are worth 1.

```
> A=1:9;P=(A>2)&(A<6)
> A
[1] 1 2 3 4 5 6 7 8 9
> P
1] FALSE FALSE TRUE TRUE FALSE FALSE FALSE
```

Return TRUE when *A* is greater than 2 and less than 6, and return FALSE otherwise.

Now we solve the second part of the problem.

```
> Q=(A>=1)&(A<6)
> Q
1] TRUE TRUE TRUE FALSE FALSE FALSE

> xor(A,Q)
1] FALSE FALSE FALSE TRUE TRUE TRUE
```

The TRUE values of the last result are the requested solution.

Chapter 3

STRUCTURES FOR STATISTICAL DATA. SEQUENCES, ARRAYS, TABLES, LISTS AND DATA FRAMES. IMPORT DATA

3.1 INTRODUCTION

This chapter addresses such essential concepts within R programming as sequences, arrays, lists and data frames, which allow a smooth work in vector and matrix analysis and all kinds of work with variables. The elements of vectors, arrays and other higher structures have assigned indexes, from which they are managed conveniently and in an orderly manner.

3.2 SEQUENCES

A sequence is simply an ordered relationship of objects. The fundamentale commands of R to work with sequences are the following:

a:b	*Returns a sequence with integers between a and b separated from unit to unit*
	
	`> S1=3:7`
	`> S1`
	`[1] 3 4 5 6 7`

seq(a,b,by=s)	*Returns the number sequence between a and b separated from s in s units*

	`> S2=seq(1,9,by=2)`
	`> S2`
	`[1] 1 3 5 7 9`

seq(a,b,s)	*Returns the number sequence between a and b separated from s in s units*
	
	`> S3=seq(1,9,2)`
	`> S3`
	`[1] 1 3 5 7 9`

seq(along=vector)	*Returns the sequence of consecutive integers starting at 1 and ending at the vector length*
	
	`> S4=seq(along=c(25,30,35,40,45))`
	`> S4`
	`[1] 1 2 3 4 5`

rep(a,n)	*Creates the sequence that repeats the element a n times*
	
	`> S5=rep(2,5)`
	`> S5`
	`[1] 2 2 2 2 2`

rep(vector,n))	*Creates the sequence that repeats the vector n times*
	
	`> S6=rep(c(1,2,3),3)`
	`> S6`
	`[1] 1 2 3 1 2 3 1 2 3`

seq(along=matrix)	*Returns the sequence of consecutive integers starting at 1 and ending at mxn, where m and n are the dimensions of the matrix*
	
	`>A=rbind(c(4,3,5),c(3,2,1),c(8,6,7)`
	`)`
	`> A`
	`      [,1] [,2] [,3]`
	`[1,]     4    3    5`
	`[2,]     3    2    1`
	`[3,]     8    6    7`
	
	`> S7=seq(along=A)`

	`> S7` `[1] 1 2 3 4 5 6 7 8 9`

rep(matrix,n))	*Creates the sequence that repeats the elements of the matrix by columns n times*
	
	`> S8=rep(A,2)`
	`> S8`
	`[1] 4 3 8 3 2 6 5 1 7 4 3 8 3 2 6 5 1 7`

seq(length=n, from=a, to=b)	*Returns the sequence of n equally spaced real numbers starting at a and ending at b*
	
	`> seq(length=9, from=1, to=5)`
	`[1] 1.0 1.5 2.0 2.5 3.0 3.5 4.0 4.5 5.0`

sequence(a)	*Create the sequence of consecutive integers starting at 1 and ending at a.*
	
	`> sequence(4)`
	`[1] 1 2 3 4`

sequence(a:b)	*Create the sequence of consecutive integers a+b that starts at 1 and ends at a and starts again at 1 and ends at b*
	
	`> sequence(4:5)`
	`[1] 1 2 3 4 1 2 3 4 5`

sequence(c(a,b))	*Create the sequence of consecutive integers a+b that starts at 1 and ends at a and starts again at 1 and ends at b*
	
	`> sequence(c(10,5))`
	`[1] 1 2 3 4 5 6 7 8 9 10 1 2 3 4 5`

gl(a,b)	*Create the sequence of consecutive integers a*b starting at 1 and ending at a and repeated each b time. It is usually used to define the levels of a*
	
	`> gl(3, 5)`
	`[1] 1 1 1 1 1 2 2 2 2 2 3 3 3 3 3`
	`Levels: 1 2 3`

gl(a,b, length=n)	*It creates the sequence of n consecutive integers that starts at 1 and ends at a and that is repeated each time b. Usually used to define factor levels*
	
	`> gl(3, 5, length=30)`
	`[1] 1 1 1 1 1 2 2 2 2 2 3 3 3 3 3 1 1 1 1 1 2 2 2 2 2 3 3 3 3 3`
	`Levels: 1 2 3`

gl(a,b, label=labelvector)	*Create the sequence of the labels defined in the label vector and repeat each one of them twice*
	
	`> gl(2, 6, label=c("Male", "Female"))`
	`1] Male Male Male Male`
	`7] Female Female Female Female`
	`Levels: Male Female`

Since sequences behave like vectors, all R commands and functions applicable to vectors are also applicable to sequences, as seen in the following examples:

```
> S1+S2
[1] 4  7 10 13 16

> S3^2
[1] 1  9 25 49 81

> log(S4)
[1] 0.0000000 0.6931472 1.0986123 1.3862944 1.6094379
> exp(S5)

[1] 7.389056 7.389056 7.389056 7.389056 7.389056
```

The ':' operator has priority over other arithmetic operators in an expression:

```
> 1:10-1
[1] 0 1 2 3 4 5 6 7 8 9

> 1:(10-1)
[1] 1 2 3 4 5 6 7 8 9
```

The *seq* function can also generate sequences of real numbers:

```
> seq(1, 5, 0.5)
[1] 1.0 1.5 2.0 2.5 3.0 3.5 4.0 4.5 5.0
```

where the first number indicates the beginning of the sequence, the second the end and the third the increment to be used to generate the sequence. It can also be used:

3.3 ARRAYS

An array is a well-ordered collection of individual elements. It is simply a list of elements each of which is associated with a positive integer called the element index, which represents the place (position) it occupies within the list. The essential thing is that each element is associated with an index, which can also be zero or negative, that fully identifies it, so that, to make changes in some element of the array, it is enough to refer to its indexes.

The arrays can be of one or several dimensions, depending on whether they have one or several sets of indexes that identify their elements. The most important commands that R enables for working with arrays are

<table>
<tr>
<td>array(data vector, dimension vector)</td>
<td>Creates the array whose data vector and dimension vector are as specified.

the following examples create two-dimensional arrays

<code>> array(1:5,c(2,3))</code>

<pre>
       [,1] [,2] [,3]
[1,]   1   3   5
[2,]   2   4   1
</pre>
<code>> array(1:5,c(2,2))</code>

<pre>
       [,1] [,2]
[1,]   1   3
[2,]   2   4
</pre>
The following example creates a three-dimensio nal array

<code>> array(1:5,c(2,3,3))</code>

<code>, , 1</code></td>
</tr>
</table>

<table>
<tr><td></td><td><pre>
      [,1] [,2] [,3]
[1,]   1   3   5
[2,]   2   4   1

, , 2

      [,1] [,2] [,3]
[1,]   2   4   1
[2,]   3   5   2

, , 3

      [,1] [,2] [,3]
[1,]   3   5   2
[2,]   4   1   3
</pre></td></tr>
<tr><td>array(data vector, dim=dimension vector))</td><td>Syntax equivalent to the previous one<pre>
> array(1:5,dim=c(2,3))
      [,1] [,2] [,3]
[1,]   1   3   5
[2,]   2   4   1

> array(1:5,dim=c(2,3,3))
, , 1

      [,1] [,2] [,3]
[1,]   1   3   5
[2,]   2   4   1

, , 2

      [,1] [,2] [,3]
[1,]   2   4   1
[2,]   3   5   2

, , 3

      [,1] [,2] [,3]
[1,]   3   5   2
[2,]   4   1   3
</pre></td></tr>
<tr><td>array(data vector, two-dimensional vector)</td><td>Creates the matrix whose elements are those specified in the data vector by columns and with the dimensions indicated in the dimension vector.</td></tr>
</table>

	`> A=array(c(9,8,9,6,1,3), c(2,3))`
	`> A`
	`     [,1] [,2] [,3]`
	`[1,]    9    9    1`
	`[2,]    8    6    3`

array(data vector, dim= two dimensional vector))	*Creates the matrix whose elements are those specified in the data vector by columns and with the dimensions indicated in the dimension vector*
	
	`> A=array(c(9,8,9,6,1,3), dim=c(2,3)`
	`)`
	`> A`
	`     [,1] [,2] [,3]`
	`[1,]    9    9    1`
	`[2,]    8    6    3`

array(data vector, dim= (1,length(data vector))	*Creates the vector whose elements are those specified in the data vector.* ***The vector is a particular case of array***
	
	`> A=array(c(9,8,9,6,1,3), dim=c(1,6)`
	`)`
	`> A`
	`     [,1] [,2] [,3] [,4] [,5] [,6]`
	`[1,]    9    8    9    6    1    3`

array(data vector, (1,length(data vector))	*Creates the vector whose elements are those specified in the data vector*
	
	`> A=array(c(9,8,9,6,1,3), c(1,6))`
	`> A`
	`     [,1] [,2] [,3] [,4] [,5] [,6]`
	`[1,]    9    8    9    6    1    3`

dim(X)	*Decreasing the dimensions of the X-array*
	
	`> A=array(1:5,dim=c(2,3,3));`
	`> dim(A)`
	`[1] 2 3 3`

aperm(array)	*Transpose an array*
	
	`> A=array(1:5,dim=c(2,3,3))`
	`> A`

```
, , 1

     [,1] [,2] [,3]
[1,]    1    3    5
[2,]    2    4    1

, , 2

     [,1] [,2] [,3]
[1,]    2    4    1
[2,]    3    5    2

, , 3

     [,1] [,2] [,3]
[1,]    3    5    2
[2,]    4    1    3

> AT=aperm(A)
> AT
, , 1

 [,1] [,2] [,3]
[1,]    1    3    5
[2,]    2    4    1
[3,]    3    5    2

, , 2

     [,1] [,2] [,3]
[1,]    2    4    1
[2,]    3    5    2
[3,]    4    1    3
```

Indexed variables can be used in arithmetic expressions and the result is an indexed variable formed from the element by element operations of the underlying vectors. The dim attributes of the operands must be generally the same and will match the dimension vector of the result. So, if A, B, and C are similar indexed variables, then

```
> D = 2*A*B + C + 1
```

stores in D a similar indexed variable, whose data vector is the result of the operations indicated on the data vectors underlying A, B and C.

The exact rules corresponding to the calculations in which indexed variables and vectors are mixed must be carefully studied.

As it has already been shown, it is important to emphasize that vectors and matrices are particular cases of arrays.

3.4 LISTS

In R, a list is an object consisting of an ordered collection of objects, known as components. The components do not have to be of the same nature, so a list can be composed of, for example, a numerical vector, a logical value, an array, and a function. This fact differentiates lists from vectors. The most important commands that R enables for working with arrays are the following:

<table>
<tr><td>list(items)</td><td>Creates the smooth with the specified elements

<code>> List=list(name="Peter", wife="Mar
y", no.children=3, age.children=c(4
,7,9))</code>

<code>> List
$name
1] "Peter"

$wife
1] "Mary"

no.children
[1] 3

$age.children
[1] 4 7 9</code></td></tr>
<tr><td>list[[i]]</td><td>Returns the 1st level i-i item on the list

<code>> List[[2]]
1] "Mary"

> List[[4]]
[1] 4 7 9</code></td></tr>
<tr><td>list[[i]][j]</td><td>Returns the second level j-item of the first level i-item in the list

<code>> List[[4]][2]
[1] 7</code></td></tr>
</table>

	The 7 is the second element of the vector (second level) which constitutes the fourth element of the first level of the list

list$element	*Returns the specified top-level item in the list without having to use its number.*
	
	`> WifeList`
	`1] "Mary"`
	`> List$name`
	`1] "Peter"`
	`> List$edad.hijos`
	`[1] 4 7 9`

list["item"]]	*Returns the specified top-level item in the list without having to use its number.*
	
	`> List ["wife"]]`
	`1] "Mary"`
	`> List ["name"]]`
	`1] "Peter"`
	`> List ["age.children"]]`
	`[1] 4 7 9`

list[i]	*Return the sublist iésima of the list*
	
	`> List [4]`
	`$age.children`
	`[1] 4 7 9`

lenght(list)	*Returns the number of first level items in the list*
	
	`> length(List)`
	`[1] 4`

list.ABC = c(list.A, list.B, list.C)	*Returns the list resulting from concatenation of the specified lists*

It is very important to distinguish clearly between `Lst[[1]]` and `Lst[1]`. `` `[. . .]' is `` the operator used to select a single element, while `[. . .]' is a general indexing operator that returns a sublist from the list. That is, `Lst[[1]] is` the first object in the `Lst list,` and if it is a list with names, the name is not included. For its part, Lst `[1], is` a sublist of the list Lst consisting of the first component. If the list has a name, it is transferred to the sublist.

The names of the components can be abbreviated to the minimum number of letters necessary to identify them exactly. Thus, in

```
> List = list(coefficients=c(1.3,4), covariance=.87)
```

`Lst$coefficients` can be specified by `List$coe,` and `List$covariance` as `List$cov`.

3.5 DATA FRAMES

The data frame structure is a list whose components must be vectors (numerical, character, logical), factors, numerical matrices, lists, or other data frames. If they are vectors, they must all be of equal length, and if they are matrices, they must all have the same number of rows.

It is common to structure a data frame as a set of column vectors of equal length. We can imagine the data frame as a rectangular matrix, whose columns are its variables and which can be of different types: numerical, factor, logical, etc.

To build a data frame type variable you can use the function *data.frame()*. Let's see an example of data frame with 2 columns, built from a numerical vector *v,* which will be a column of name value and another vector that will be another column of name *case* formed with qualitative data.

```
> v=c(2.1,1.4,6,3.5,8);v

[1] 2.1 1.4 6.0 3.5 8.0

>
DatosEstruc=data.frame(value=v,case=c("high","low","medium","
high","medium"))

> StructureData

 case value
1    2.1 high
2    1.4 low
3        medium
4    3.5 high
5    8.0 average
```

We refer separately to each column by separating with the $ sign the name of the data frame and the name of the column (which is like the name of the associated variable).

```
> DatosEstruc$valor;DatosEstruc$caso

[1] 2.1 1.4 6.0 3.5 8.0
1] high low medium high medium
Levels: high low medium
```

To see the dimension of an object, in this case the data frame DatosEstruc, you can use the *dim* function:

```
> dim(StructureData)

[1] 5 2
That is, it has 2 columns, and each one has 5 rows.
```

The R summary() function applied to the data frame gives us the statistical summary of its variables (columns)

```
> summary(DatosEstruc)
      case value
 Min.    :1.4    high :2
 1st Qu. :2.1    under :1
 Median  :3.5    medium:2
 Mean    :4.2
 3rd Qu. :6.0
 Max.    :8.0
```

Let's see some instructions to access the information in the data frame. You can understand that you work with the data.frame as if it was an array with rows and columns:

```
> StructureData[1,2]
[1] stop
Levels: high low medium

> DataEstruc[0,]
1] case value
<0 rows> (or 0-length row.names)

> DataStruc[1,]
 case value
   1   2.1 high

> StructureData [1:3,]
 case value
```

```
1     2.1 high
2     1.4 low
3     6.0 medium

> StructureData [1:3,2]
1] high low medium
Levels: high low medium

> new_order=c(2,1,4,3,5);new_order
[1] 2 1 4 3 5

> new_order=c(2,1,4,3,5);new_order
[1] 2 1 4 3 5

> DatosEstruc_ordenado=DatosEstruc [new_order,];
> DataStructure_Orderings
 case value
2     1.4 low
1     2.1 high
4     3.5 high
3     6.0 medium
5     8.0 average
```

Let's see how to generate from the R Console environment a vector of random nameNormal with random numbers coming from a normal distribution of mean 2 and standard deviation 4. We calculate the mean and standard deviation of the sample using different R functions:

```
> aleatNormal=rnorm(15,2,4);aleatNormal

 [1] -1.919925930 0.880112011 2.915664570 -1.827781189 0.371865595
 [6] 1.737658371 1.299319151 -0.747209025 5.202930682 -2.966065683
[11] 1.353941243 0.002939512 0.946853283 -5.835235018 3.582367120

> mean(aleatNormal);sd(aleatNormal)
[1] 0.3331623
[1] 2.745425
> summary(aleatNormal)
    Min. 1st Qu. Median   Mean  3rd Qu.   Max.
-5.8350 -1.2870 0.8801    0.3332 1.5460 5.2030
> numSummary(aleatNormal)

 mean         sd          0%         25%        50%       75%      n
0.3331623 2.745425 -5.835235 -1.287495 0.880112 1.545800 5.202931 15
```

The most important commands that R enables for working with data frames are the following:

A=data.frame(x,y,z,..)	*It creates a data frame whose elements are the columns x, y, z,.*
	
	`> datos1=data.frame(Peso=c(90,120,66), A`
	`ltura=c(1.90,1.87,1.70))`
	`> data1`
	`  Weight Height`
	`1     90     1.90`
	`2    120     1.87`
	`3     66     1.70`

B=data.frame(A,x,y,..)	*Creates a data frame containing other data frames as elements (A is a data frame) and other additional components (x,y,..). You can nest dataframes inside other dataframes*
	
	`> dato2=data.frame(dato1,Sexo=c("Hombre`
	`", "Hombre", "Mujer"))`
	`> data2`
	`  Weight Height Sex`
	`1     90     1.90 Man`
	`2    120     1.87 Male`
	`3     66     1.70 Female`

A[n]	*Returns the nth element of data frame A*
	
	`> data2[3]`
	`     Sex`
	`1 Man`
	`2 Man`
	`3 Women`

A$element	*It extracts the specified element from dataframe A and transforms it into a vector.*
	
	`> Weight=data2$Weight`
	`> Weight`
	`[1] 90 120 66`
	
	This is important because from this moment on the element extracted from the data frame can have all the commands and functions related to vectors

I( )	*Allows to enter a vector in a data frame.*

<table>
<tr><td></td><td><pre>
> names=c("Pepe","Paco","Pepita")
> names
1] "Pepe"   "Paco"   "Pepita"
> datos3=data.frame(datos2, Names=I(nam
es))
> data3
 Weight Height   Names Sex
1   90   1.90   Pepe Man
2   120   1.87   Paco Man
3   66   1.70   Pepita Woman
</pre></td></tr>
<tr><td>subset(A,select=element)</td><td>Extracts the specified element from the dataframe.<pre>
> data3
 Weight Height   Names Sex
1   90   1.90   Pepe Man
2   120   1.87   Paco Man
3   66   1.70   Pepita Woman
> subset(data3,select=c(Gender,Names))
     Names Sex
1   Pepe Man
2   Paco Man
3   Nugget Woman
</pre></td></tr>
<tr><td>transform(A, transformation)</td><td>Performs the specified transformation on an element of data frame A<pre>
> transform(data3,logPeso=log(Weight))
 Weight Height   Gender Names logWeigh
t
1   90   1.90 Man   Pepe 4.499810
2   120   1.87 Male   Paco 4.787492
3   66   1.70 Pepita   Woman 4.189655

> transform(data3,BMI=Weight/(Height)^2
)

 Weight Height   Gender   BMI Name
s
1   90   1.90 Man   Pepe 24.93075
2   120   1.87 Male   Paco 34.31611
3   66   1.70 Woman   Pepita 22.83737
</pre></td></tr>
<tr><td>dim(A)</td><td>Returns data frame dimension A</td></tr>
</table>

<table>
<tr><td></td><td><pre>> dim(data3)
[1] 3 4</pre></td></tr>
<tr><td>attach(A)</td><td>Enables data frame A elements to be accessible as vectors without using the A$element command<pre>> attach(data3)
The following objects are masked _by_ .
GlobalEnv:

 Name, Weight, Sex

> Weight
[1] 90 120 66
> sum(Weight)
[1] 276</pre></td></tr>
<tr><td>dettach(A)</td><td>Disable data frame A elements to be accessible as vectors without using the A$element command</td></tr>
<tr><td>expand.grid(x,y,..)</td><td>Create a dataframe with all combinations of the values of x,y,.<pre>> data4=expand.grid(c(1,2,3),c(10,20))
> data4
 Var1 Var2
1   1   10
2   2   10
3   3   10
4   1   20
5   2   20
6   3   20

> data5=expand.grid(a=c(60,80), p=c(100
, 300), sex=c("Male", "Female")

> data5
  a   p sex
1 60 100 Male
2 80 100 Male
3 60 300 Male
4 80 300 Male
5 60 100 Female
6 80 100 Female
7 60 300 Female
8 80 300 Female</pre></td></tr>
</table>

<table>
<tr><td>summarize(A)</td><td>Find descriptive statistics for all column variables of dataframe A

<pre>
> summary(data4)
       Var1   Var2
 Min.   1.00   Min.   :10
 1st Qu.:1.25   1st Qu. :10
 Median :2.00   Median :15
 Mean   :2.00   Mean   :15
 3rd Qu.:2.75   3rd Qu. :20
 Max.   :3.00   Max.   :20
</pre></td></tr>
</table>

3.6 IMPORT OF STATISTICAL DATA AS DATA FRAMES

3.6.1 Text file import

We already know that R supports direct import of ASCII text files. The syntax for importing is as follows:

```
mydata=read.table("c:/mydata.txt", header=TRUE, sep=",")
```

The previous syntax imports the file named mydata.txt located in C:/ keeping the headers of the file as names of its variables and considering that the separator between elements is the comma. However, if the file is columnized (variables as clummas), it is not necessary to specify the separator.

When an ASCII file is imported, the R language considers it with a data frame structure. That means that only commands and functions related to data frames can be applied. Let's see an example:

```
> data = read.table("H:/R/data/zonasmadR.txt", header=TRUE)
> data
             b    pt  p14  p65   p10 anal  nes  ocu ocuin ocuser  tec  pd   tm
1       Centro 166.5 23.3 38.1 152.8  4.2 21.4 54.1   7.6   41.7  8.8 0.8 10.3
2   Arganzuela 121.1 23.5 18.4 106.1  2.0 16.5 69.4   7.6   28.6  7.2 0.6  8.4
3       Retiro 126.0 27.2 16.8 109.2  1.2 28.1 39.9   6.3   30.1 10.4 1.9  4.7
4    Salamanca 180.0 30.5 33.4 162.1  0.5 45.3 57.5   7.6   45.1 16.1 2.6  5.4
5    Chamartín 180.0 30.5 16.1 130.3  1.3 39.3 48.1   7.2   35.8 14.5 2.8  4.8
6       Tetuán 164.2 31.3 23.5 145.1  4.2 24.2 52.3   9.6   37.9  9.6 1.1 12.2
7     Chamberi 182.7 29.4 35.0 165.4  1.8 47.2 59.4   7.5   46.4 17.1 2.4  6.1
8   Fuencarral 176.2 51.3 15.6 142.2  3.6 21.6 95.6  10.3   38.7 11.1 1.6 14.3
9      Moncloa 108.4 23.4 13.4  94.2  1.5 32.5 34.6   5.3   26.0  8.7 1.4  5.5
10      Latina 289.5 79.5 23.1 239.7  6.0 22.7 86.6  17.7   59.8 10.4 1.3 26.4
11 Carabanchel 255.9 60.5 24.1 218.3  7.3 16.6 77.4  19.4   50.1  7.5 1.0 28.2
12  Villaverde 195.0 48.5 16.1 166.1  8.3  9.0 56.6  19.3   30.7  3.7 0.5 26.3
13    Mediodía 171.7 49.3 11.1 139.9  9.8  5.5 48.5  13.3   28.1  2.9 0.3 22.9
14    Vallecas 186.2 42.2 20.3 159.8 10.3  7.2 53.7  13.6   32.5  3.1 0.3 23.8
15   Moratalaz 145.9 40.8 10.9 121.4  3.9 10.1 73.7  16.5   49.0 11.6 2.1 19.7
16 CiudadLineal 135.1 55.3 21.9 201.5 4.3 28.2 73.7  16.5   49.0 11.6 2.1 19.7
17     SanBlas 137.7 32.1 10.3 118.5  6.0  6.3 41.4  12.2   24.1  2.9 0.2 18.2
18   Hortaleza 167.9 51.4 10.1 132.6  4.0 15.5 51.6  12.3   33.7  7.9 1.4 15.8
```

```
> class(data)
1] "data.frame"
```

We can apply the summary function to the data frame.

```
> summary(data)

               b              pt              p14              p65              p10
 Arganzuela   : 1   Min.   :108.4   Min.   :23.30   Min.   :10.10   Min.   : 94.2
 Carabanchel  : 1   1st Qu.:139.8   1st Qu.:29.68   1st Qu.:13.95   1st Qu.:123.6
 Centro       : 1   Median :169.8   Median :36.45   Median :17.60   Median :143.7
 CiudadLineal: 1    Mean   :171.7   Mean   :40.56   Mean   :19.90   Mean   :150.3
 Chamartín    : 1   3rd Qu.:182.0   3rd Qu.:50.80   3rd Qu.:23.40   3rd Qu.:164.6
 Chamberi     : 1   Max.   :289.5   Max.   :79.50   Max.   :38.10   Max.   :239.7
 (Other)      :12
      anal               nes              ocu              ocuin            ocuser
 Min.   : 0.500   Min.   : 5.50   Min.   :34.60   Min.   : 5.30   Min.   :24.10
 1st Qu.: 1.850   1st Qu.:11.45   1st Qu.:49.27   1st Qu.: 7.60   1st Qu.:30.25
 Median : 4.100   Median :21.50   Median :55.35   Median :11.25   Median :36.85
 Mean   : 4.456   Mean   :22.07   Mean   :59.67   Mean   :11.66   Mean   :38.18
 3rd Qu.: 6.000   3rd Qu.:28.18   3rd Qu.:72.62   3rd Qu.:15.78   3rd Qu.:46.08
 Max.   :10.300   Max.   :47.20   Max.   :95.60   Max.   :19.40   Max.   :59.80

      tec               pd               tm
 Min.   : 2.900   Min.   :0.200   Min.   : 4.700
 1st Qu.: 7.275   1st Qu.:0.650   1st Qu.: 6.675
 Median : 9.200   Median :1.350   Median :15.050
 Mean   : 9.172   Mean   :1.356   Mean   :15.150
 3rd Qu.:11.475   3rd Qu.:2.050   3rd Qu.:22.100
 Max.   :17.100   Max.   :2.800   Max.   :28.200
```

However, we cannot use your individual variables as vectors to work with. If we try to add the pt elements we get an error.

```
> sum(pt)
Error in sum(pt) : invalid 'type' (closure) of argument
```

To solve this problem, it is enough to define the variables of the data frame as vectors.

```
> pt=data$pt
> sum(pt)
[1] 3090
```

But the previous task would have to be carried out variable by variable. However, as we have already seen, all the columns of the data frame can be transformed into vectors with the *attach* function.

```
> attach(data);
```

Now we can use any column of the data frame as a vector.

```
> sum(ocuin)
```

```
[1] 209.8
```

The import of the ASCII file can also be done from the RStudio environment through the option *Import Dataset → From Text File*. In the lower area of the following figure it can be seen that the import is done with Data Frame structure. When you click on *Import,* the import is done.

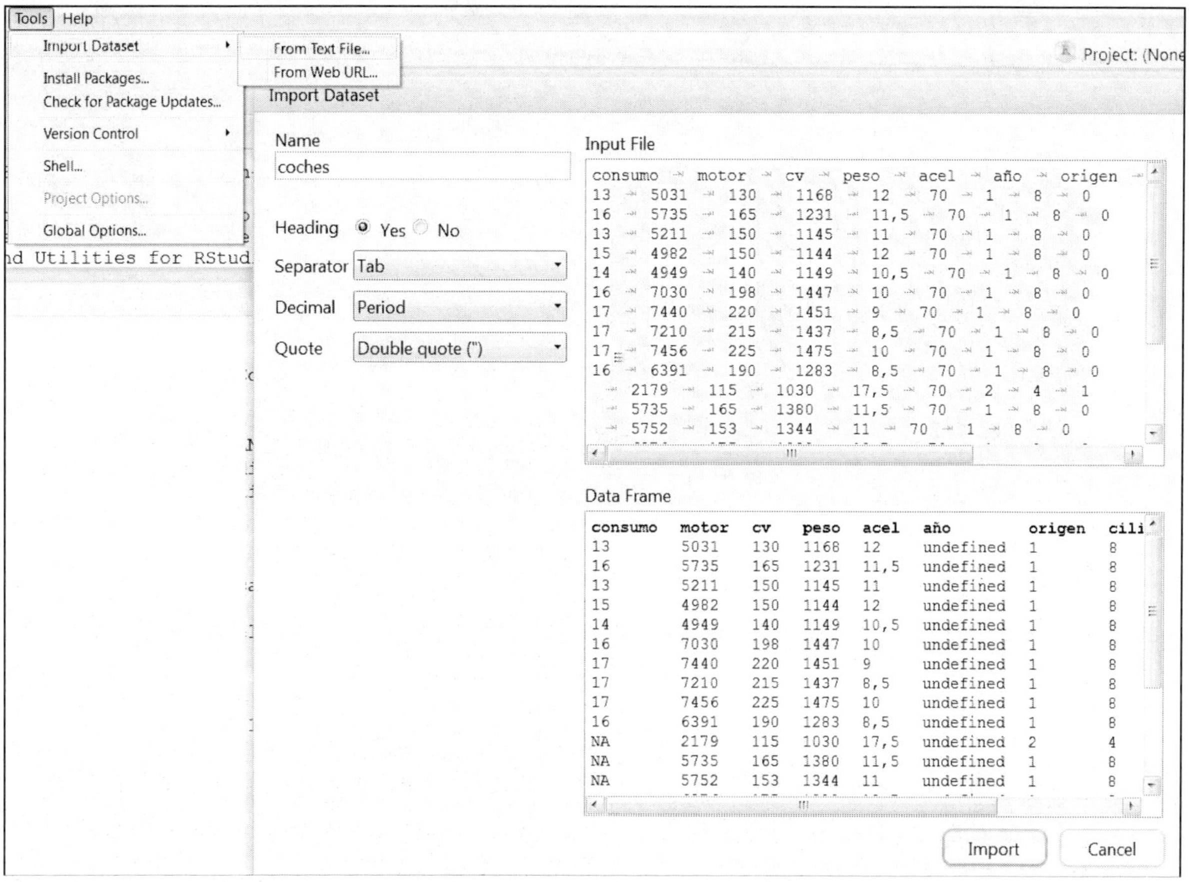

3.6.2 Import of Excel files

To import Excel files it is necessary to install and preload the *xlsx* library.

The following syntax is used to install the *xlsx* library:

```
> install.packages("xlsx")
```

You can also use the *Install packages* option from the RStudio *Tools* menu, as shown in the figure below:

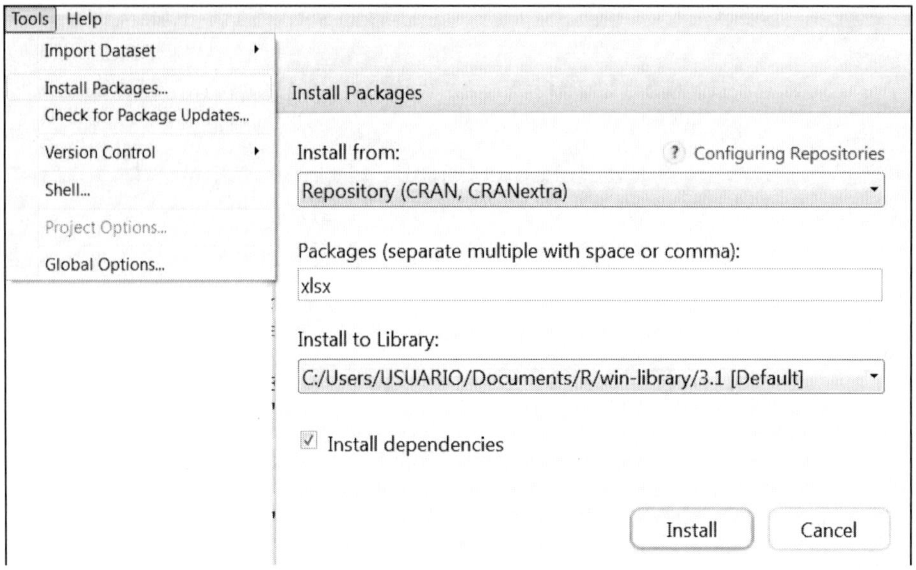

You can also use the *Install Packages* option from the *RConsole Packages* menu.

Once the library is installed, it is enabled through the syntax:

```
> library(xlsx)
```

The following syntax is used to import the file *zonasmad.xls* located in the path *H:\R|data:*

```
> dataE = read.xlsx("H:/R/data/zonasmad.xls",1)
```

With the above syntax the first sheet of the Excel file is read. To read any other sheet the following syntax is used:

```
> dataE = read.xlsx("H:/R/data/zonasmad.xls",sheetName = "shee
tname"))
```

3.6.3 Importing SPSS files

To import IBM SPSS files keeping labels and their values it is necessary to install and load previously the *foreing* library. The syntax is:

```
> install(foreign)
> library(foreign)
> dataS=read.spss("H:/R/data/zonasmad.sav", use.value.labels =
TRUE)
> dataS
$b
```

```
 [1] "Centro             " "Arganzuela         " "Retiro             "
 [4] "Salamanca          " "Chamartín          " "Tetuán             "
 [7] "Chamberi           " "Fuencarral         " "Moncloa            "
[10] "Latina             " "Carabanchel        " "Villaverde         "
[13] "Mediodía           " "Vallecas           " "Moratalaz          "
[16] "Ciudad Lineal      " "San Blas           " "Hortaleza          "

$pt
 [1] 166.5 121.1 126.0 180.0 180.0 164.2 182.7 176.2 108.4 289
.5 255.9 195.0 171.7
[14] 186.2 145.9 135.1 137.7 167.9

$p14
 [1] 23.3 23.5 27.2 30.5 30.5 31.3 29.4 51.3 23.4 79.5 60.5 48
.5 49.3 42.2 40.8
[16] 55.3 32.1 51.4

$p65
 [1] 38.1 18.4 16.8 33.4 16.1 23.5 35.0 15.6 13.4 23.1 24.1 16
.1 11.1 20.3 10.9
[16] 21.9 10.3 10.1

$p10
 [1] 152.8 106.1 109.2 162.1 130.3 145.1 165.4 142.2  94.2 239
.7 218.3 166.1 139.9
[14] 159.8 121.4 201.5 118.5 132.6

$anal
 [1]  4.2  2.0  1.2   NA  1.3  4.2  1.8  3.6  1.5  6.0  7.3  8
.3  9.8 10.3  3.9
[16]  4.3  6.0  4.0

$nes
 [1] 21.4 16.5 28.1 45.3 39.3 24.2 47.2 21.6 32.5 22.7 16.6  9
.0  5.5  7.2 10.1
[16] 28.2  6.3 15.5

$ocu
 [1] 54.1 69.4 39.9 57.5 48.1 52.3 59.4 95.6 34.6 86.6 77.4 56
.6 48.5 53.7 73.7
[16] 73.7 41.4 51.6

$ocuin
 [1]  7.6  7.6  6.3  7.6  7.2  9.6  7.5 10.3  5.3 17.7 19.4 19
.3 13.3 13.6 16.5
[16] 16.5 12.2 12.3

$ocuser
```

```
 [1] 41.7 28.6 30.1 45.1 35.8 37.9 46.4 38.7 26.0 59.8 50.1 30
.7 28.1 32.5 49.0
[16] 49.0 24.1 33.7

$tec
 [1]  8.8  7.2 10.4 16.1 14.5  9.6 17.1 11.1  8.7 10.4  7.5  3
.7  2.9  3.1 11.6
[16] 11.6  2.9  7.9

$pd
 [1] 0.8 0.6 1.9 2.6 2.8 1.1 2.4 1.6 1.4 1.3 1.0 0.5 0.3 0.3 2
.1 2.1 0.2 1.4

$tm
 [1] 10.3  8.4  4.7  5.4  4.8 12.2  6.1 14.3  5.5 26.4 28.2 26
.3 22.9 23.8 19.7
[16] 19.7 18.2 15.8
```

We check that the import has been done as a list.

```
> class(dataS)
1] "list"
```

To access the elements in the list (variables in the file) as vectors, it is necessary to use the $ variable by variable operator.

```
> pt=dataS$pt
> pt
 [1] 166.5 121.1 126.0 180.0 180.0 164.2 182.7 176.2 108.4 289
.5 255.9 195.0 171.7
[14] 186.2 145.9 135.1 137.7 167.9
> sum(pt)
[1] 3090
```

We can also use the attach function to simultaneously enable all variables in the file as vectors.

```
> attach(datosS)

> sum(ocu)
[1] 1074.1
```

We could also have imported the SPSS file as a data frame using the following syntax:

```
> datosS=read.spss("H:/R/data/zonasmad.sav", use.value.labels
= TRUE, to.data.frame = TRUE)
```

3.6.4 Importing STATA files

To import STATA files with labels and their values, the *foreing* library must be installed and loaded beforehand. The syntax is:

```
> install(foreign)
> library(foreign)
> datosSTATA=read.dta("H:/R/data/zonasmad.dta")
> datosSTATA
                b    pt  p14  p65   p10 anal  nes  ocu ocuin ocuser  tec  pd   tm
1          Centro 166.5 23.3 38.1 152.8  4.2 21.4 54.1   7.6   41.7  8.8 0.8 10.3
2      Arganzuela 121.1 23.5 18.4 106.1  2.0 16.5 69.4   7.6   28.6  7.2 0.6  8.4
3          Retiro 126.0 27.2 16.8 109.2  1.2 28.1 39.9   6.3   30.1 10.4 1.9  4.7
4       Salamanca 180.0 30.5 33.4 162.1   NA 45.3 57.5   7.6   45.1 16.1 2.6  5.4
5       Chamartín 180.0 30.5 16.1 130.3  1.3 39.3 48.1   7.2   35.8 14.5 2.8  4.8
6          Tetuán 164.2 31.3 23.5 145.1  4.2 24.2 52.3   9.6   37.9  9.6 1.1 12.2
7        Chamberi 182.7 29.4 35.0 165.4  1.8 47.2 59.4   7.5   46.4 17.1 2.4  6.1
8      Fuencarral 176.2 51.3 15.6 142.2  3.6 21.6 95.6  10.3   38.7 11.1 1.6 14.3
9        Moncloa 108.4 23.4 13.4  94.2  1.5 32.5 34.6   5.3   26.0  8.7 1.4  5.5
10         Latina 289.5 79.5 23.1 239.7  6.0 22.7 86.6  17.7   59.8 10.4 1.3 26.4
11    Carabanchel 255.9 60.5 24.1 218.3  7.3 16.6 77.4  19.4   50.1  7.5 1.0 28.2
12     Villaverde 195.0 48.5 16.1 166.1  8.3  9.0 56.6  19.3   30.7  3.7 0.5 26.3
13       Mediodía 171.7 49.3 11.1 139.9  9.8  5.5 48.5  13.3   28.1  2.9 0.3 22.9
14       Vallecas 186.2 42.2 20.3 159.8 10.3  7.2 53.7  13.6   32.5  3.1 0.3 23.8
15     Moratalaz 145.9 40.8 10.9 121.4  3.9 10.1 73.7  16.5   49.0 11.6 2.1 19.7
16 Ciudad Lineal 135.1 55.3 21.9 201.5  4.3 28.2 73.7  16.5   49.0 11.6 2.1 19.7
17       San Blas 137.7 32.1 10.3 118.5  6.0  6.3 41.4  12.2   24.1  2.9 0.2 18.2
18      Hortaleza 167.9 51.4 10.1 132.6  4.0 15.5 51.6  12.3   33.7  7.9 1.4 15.8
```

In this case the import is done in vector mode, so that all variables are already accessible as vectors.

```
> pt
 [1] 166.5 121.1 126.0 180.0 180.0 164.2 182.7 176.2 108.4 289
.5 255.9 195.0 171.7
[14] 186.2 145.9 135.1 137.7 167.9
> sum(pt)
[1] 3090
```

3.6.5 SAS File Deployment

To import SAS files it is necessary to previously install and load the *foreing* library. The syntax is:

```
> install(foreign)
> library(foreign)
> datosSAS=read.ssd("H:/R/data/", "zonasmad")
```

3.7 TIME SERIES

The function *ts* creates a class object "ts" (time series) from a vector (inivariate time series) or a matrix (multivariate series). The syntax and options of the *ts* function are presented below:

```
ts(data = NA, start = 1, end = numeric(0), frequency = 1,
deltat = 1, class, names)
```

The options have the following interpretation:

`data` a vector or a matrix

`start` the time of the first observation either a number or a vector with two integers (see example below)

`end` the time of the last observation specified in the same way as `start`

`frequency` the number of observations per time unit

`deltat` the fraction of the sampling period between successive observations (e.g. `1/12` for monthly data); you only have to specify or `frequency` or `deltat`

`class class to` be assigned to the object; default is `"ts"` for a series univariate, and `c("mts", "ts")` for a multivariate series

`names` for a multivariate series, a character type vector with the names of the individual series; by default the names of the `data,` or `Series 1, Series 2, .`

Below are some examples with `ts()`:

```
> ts(1:10, start = 1959)
Time Series:
Start = 1959
End = 1968
Frequency = 1
[1] 1 2 3 4 5 6 7 8 9 10

> ts(1:47, frequency = 12, start = c(1959, 2))
Jan Feb Mar Apr May Jun Jul Aug Sep Oct Nov Dec
1959 1 2 3 4 5 6 7 8 9 10 11
1960 12 13 14 15 16 17 18 19 20 21 22 23
```

```
1961 24 25 26 27 28 29 30 31 32 33 34 35
1962 36 37 38 39 40 41 42 43 44 45 46 47

> ts(1:10, frequency = 4, start = c(1959, 2))
Qtr1 Qtr2 Qtr3 Qtr4
1959 1 2 3
1960 4 5 6 7
1961 8 9 10

> ts(matrix(rpois(36, 5), 12, 3), start=c(1961, 1),
frequency=12)
Series 1 Series 2 Series 3
Jan 1961 8 5 4
20
Feb 1961 6 6 9
Mar 1961 2 3 3
Apr 1961 8 5 4
May 1961 4 9 3
June 1961 4 6 13
Jul 1961 4 2 6
Aug 1961 11 6 4
Sep 1961 6 5 7
Oct 1961 6 5 7
Nov 1961 5 5 7
Dec 1961 8 5 2
```

Exercise 1. We consider the sequence of integers from zero to six. Find the sequence that contains its squares.

```
> S1=seq(1:6)
> S1
[1] 1 2 3 4 5 6
> S2=seq(1:6)^2
> S2
[1] 1  4 9 16 25 36
```

Exercise 2. Calculate: $\sum ne^{-n}$ $\prod n^2 2^n/e^n$ *(n=1..10)*

```
> sum((1:10)*exp(1:10)^(-(1:10)))
[1] 0.4048814

> prod((1:10)^2*2^(1:10)/exp(1)^(1:10))
[1] 616565.6
```

Exercise 3. Calculat e: $\sum_{k=1}^{100} k^3 \sum_{k=1}^{100} k^4 \ y \sum_{k=1}^{100} k^5$

```
> sum(1:100)^3
[1] 128787625000

> options(digits=20)
> sum(1:100)^4
[1] 650377506250000.00000

> sum(1:100)^5
[1] 3284406406562500096
```

Exercise 4. Calculate the arithmetic mean and the geometric mean of the first 1000 positive integers

```
> M=sum(1:100)/100
> M
[1] 50.5

> G=prod(1:100)^(1/100)
> G
[1] 37.99269
```

Exercise 5. Let's consider a three-dimensional array whose elements are three square matrices of order 3 formed by the first 27 positive integers. Define the array and find the inverse matrix of the product of its elements. Find also the self values and self vectors of the sum of its elements.

We begin by properly defining the three-dimensional array.

```
> A=array(1:27, dim=c(3,3,3))


> A
, , 1

     [,1] [,2] [,3]
[1,]    1    4    7
[2,]    2    5    8
[3,]    3    6    9

, , 2
```

```
     [,1] [,2] [,3]
[1,]   10   13   16
[2,]   11   14   17
[3,]   12   15   18

, , 3

     [,1] [,2] [,3]
[1,]   19   22   25
[2,]   20   23   26
[3,]   21   24   27
```

Then we extract its elements (three square matrices of order 3).

```
> M1=A[,,1]
> M1
     [,1] [,2] [,3]
[1,]    1    4    7
[2,]    2    5    8
[3,]    3    6    9

> M2=A[,,2]
> M2
     [,1] [,2] [,3]
[1,]   10   13   16
[2,]   11   14   17
[3,]   12   15   18

> M3=A[,,3]
> M3
     [,1] [,2] [,3]
[1,]   19   22   25
[2,]   20   23   26
[3,]   21   24   27
```

We now calculate the inverse product matrix of the three elements.

```
> solve(M1*M2*M3)
            [,1]       [,2]        [,3]
[1,] 0.16393199 -0.2873789 0.12738046
[2,] -0.20605196 0.3538731 -0.15417236
[3,] 0.07342013 -0.1250817 0.05434675
```

Finally, we calculate the values and vectors of the sum of the three elements.

```
> eigen(M1+M2+M3)
$values
```

```
1] 1.272729e+02 -1.272856e+00 -6.997266e-16

$vectors
           [,1]        [,2]       [,3]
[1,] -0.5356313 -0.79526971 0.4082483
[2,] -0.5763903 -0.09853771 -0.8164966
[3,] -0.6171493 0.59819429 0.4082483
```

Exercice 6. Generate a three-dimensional array that contains the first eight natural non-negative numbers and whose three indexes vary from 1 to 2. Generate the array whose elements are the squares of the previous array and add them

```
> A=array(1:8, dim=c(2,2,2))
> A
, , 1

     [,1] [,2]
[1,]    1    3
[2,]    2    4

, , 2

     [,1] [,2]
[1,]    5    7
[2,]    6    8

> A1=A[,,1]
> A1
     [,1] [,2]
[1,]    1    3
[2,]    2    4

> A2=A[,,2]

> A2
     [,1] [,2]
[1,]    5    7
[2,]    6    8

> S=A1+A2
> S
     [,1] [,2]
[1,]    6   10
[2,]    8   12

> B=A^2
> B
```

```
, , 1

     [,1] [,2]
[1,]    1    9
[2,]    4   16

, , 2

     [,1] [,2]
[1,]   25   49
[2,]   36   64

> B1=B[,,1]
> B1
     [,1] [,2]
[1,]    1    9
[2,]    4   16

> B2=B[,,2]
> B2
     [,1] [,2]
[1,]   25   49
[2,]   36   64

> S2=B1+B2
> S2
     [,1] [,2]
[1,]   26   58
[2,]   40   80
```

Exercise 7. Below is a table with the students in a class, their sizes and weights. The aim is to calculate the average weight and height of the students in the class.

Student	Height	Weight
Perez	1.81	85
Gonzalez	1.70	72
Rodriguez	1.75	80
Alonso	1.90	95
Lopez	1.83	85
Label	1.75	72
Dominguez	1.81	84

We will start by assigning the information to an R data frame.

```
> Altura=c(1.81,1.70,1.75,1.90,1.83,1.75,1.81)
```

```
> Weight=c(85,72,80,95,85,72,84)
> Data=data.frame(Student,Height,Weight)
> Facts and figures
      Student Height Weight
1       Perez   1.81     85
2    Gonzalez   1.70     72
3   Rodriguez   1.75     80
4      Alonso   1.90     95
5       Lopez   1.83     85
6       Stamp   1.75     72
7   Dominguez   1.81     84
```

Next we enable the variables of the data frame so that they can be treated as vectors.

```
> attach(Data)
The following objects are masked _by_ .GlobalEnv:

 Height, Student, Weight
```

Finally we calculate the average height and average weight.

```
> MEDIUM HEIGHT=sum(Height)/length(Height)
> ALTURAMEDIA
[1] 1.792857

> AVERAGE=sum(Weight)/length(Weight)
> PESOMEDIUM
[1] 81.85714
```

Chapter 4

PREDEFINED AND CUSTOM-MADE NUMERICAL, VECTOR AND MATRIX FUNCTIONS. APPLICATIONS IN STATISTICS

4.1 FUNCTIONS

The concept of function, as well as the concept of variable, is fundamental when working with mathematical software. Obviously, the theoretical concept of mathematical function is fixed and independent of the software package. But the way to implement and handle the functions is very characteristic of each particular program. R allows to define and handle the functions in a very simple way. *In turn, R* incorporates an extensive set of functions, which we will address throughout the chapters.

4.2 DEFINITION OF FUNCTIONS

We have already seen that R incorporates multiple functions for use in daily work with the program. But, in addition, the program offers the possibility to define custom functions. The syntax to define a function is as follows:

FunctionName = function(argument1, argument2,...,argumenton)
{
sentences
return(object)
}

The returned object can be any type of data (vector, array, list, factor, data frame, etc.)

4.3 THE EDITOR OF R

The functions can be written on the R command window, but that is uncomfortable, especially for possible error correction. For this reason, the R editor is used, which allows you to write the program in text mode, with ease to make corrections in case of errors.

The R editor is invoked from RStudio via *File → New File → R Script, as shown* in Figure 4-1.

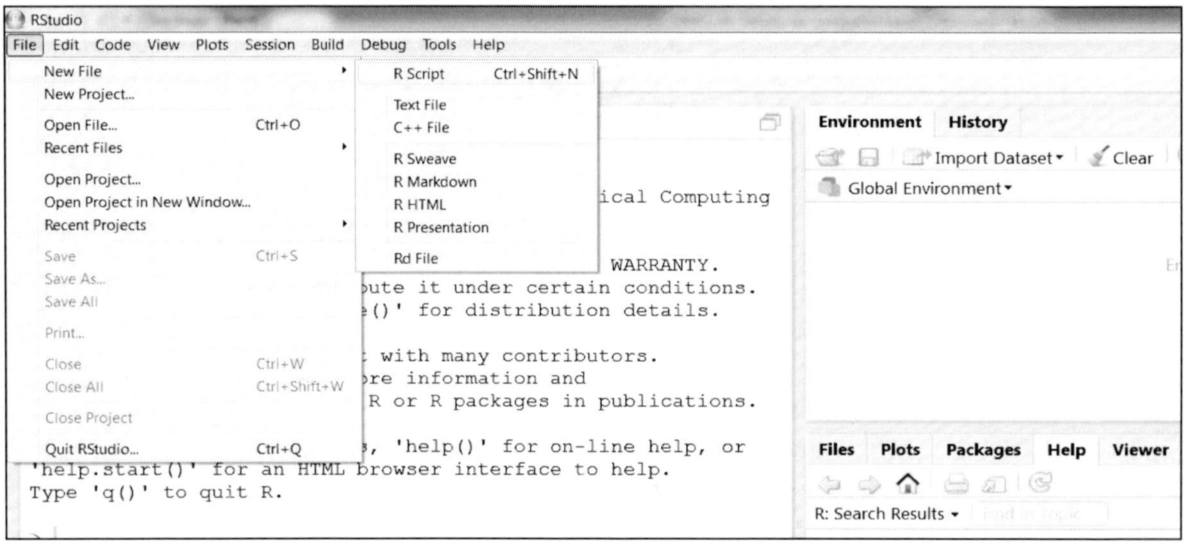

Figure 4-1

You get the editor (Figure 4-2), to be able to write the syntax of functions and programs. The *Run* button allows the execution of the syntax and the button allows to save the syntax.

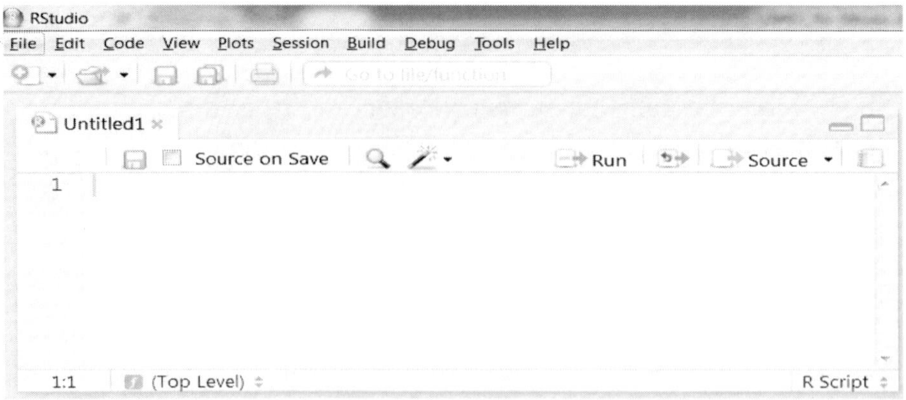

Figure 4-2

As an example we are going to construct a function that performs the sum of any two numbers. We write the syntax in the editor, select it and click on the *Run* button (Figure 4-3).

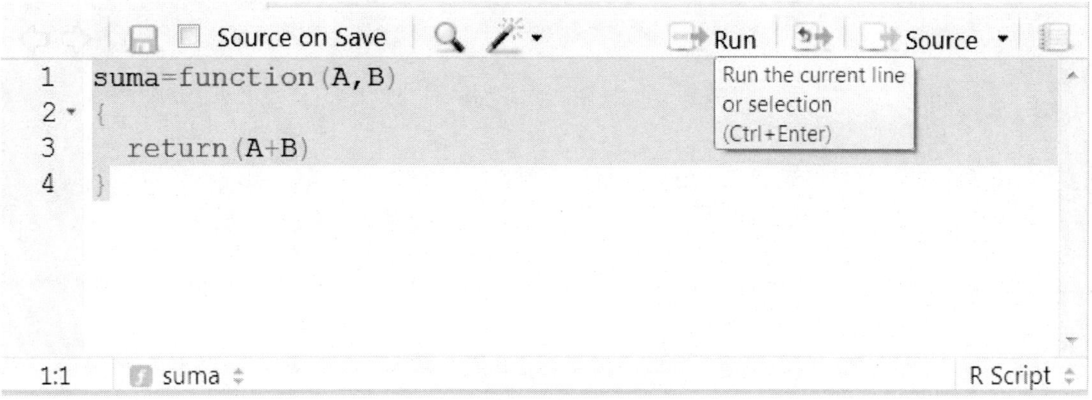

Figure 4-3

On Rconsole (R commands window) we see reflected the syntax that we have executed from the editor (Figure 4-4). If no errors appear in the console, the program syntax is correct.

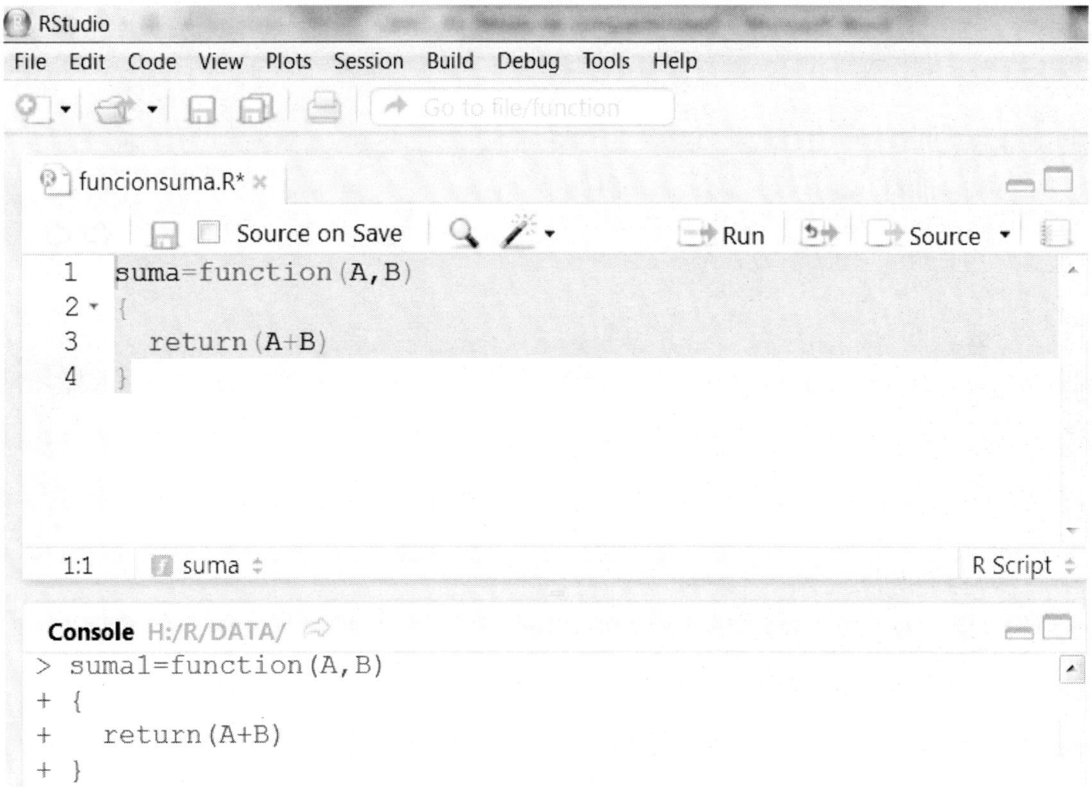

Figure 4-4

Now we can execute the function from the R command window.

```
> sum(3,8)
[1] 11
> sum(-1,75)
[1] 74
```

Since no restrictions have been placed on the function's arguments, it would also be possible to add objects of any type, for example vectors or arrays. Let's see examples:

```
> sum(c(1,2,3),c(4,5,6))
[1] 5 7 9

> A=matrix(c(1,2,3,4,5,6),3,2)
> A
     [,1] [,2]
[1,]    1    4
[2,]    2    5
[3,]    3    6

> B=matrix(c(11,12,13,14,15,16),3,2)
> B
     [,1] [,2]
[1,]   11   14
[2,]   12   15
[3,]   13   16

> sum(A,B)
     [,1] [,2]
[1,]   12   18
[2,]   14   20
[3,]   16   22
```

With the button *Save* we save the file. As it is an R program, the file will be saved with the .R extension in the default working subdirectory.

With the option *File* → *Open File* the file will be opened again in the editor, if necessary.

Then we will define the function *fun1(x)=x^3-2x+cos(x)*, creating the corresponding *file with* the name *fun1*.

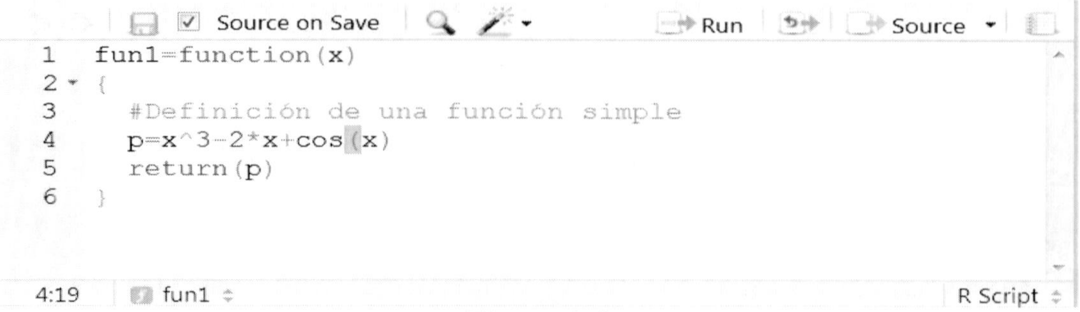

Figure 4-5

Now we can execute the function.

```
> fun1=function(x)
+ {
+     #Definition of a simple function
+     p=x^3-2*x+cos(x)
+     return(p)
+ }

> fun1(pi)
[1] 23.72309

> fun1(c(pi,2*pi,3*pi,4*pi))
[1]    23.72309 236.48384 817.31991 1960.26897
```

The following example (Figure 4-6) illustrates the definition of a function that solves a second degree equation.

It is noted that this function has three input parameters and two output parameters. The three input parameters are simply defined as the three arguments of the function. The two output parameters are defined as arguments to the return command, but in vector mode, since the return command does not support more than one parameter.

We save the *file* with the name *rootquad.R*

The function, defined in this way will allow to solve in R any second degree equation with real solutions.

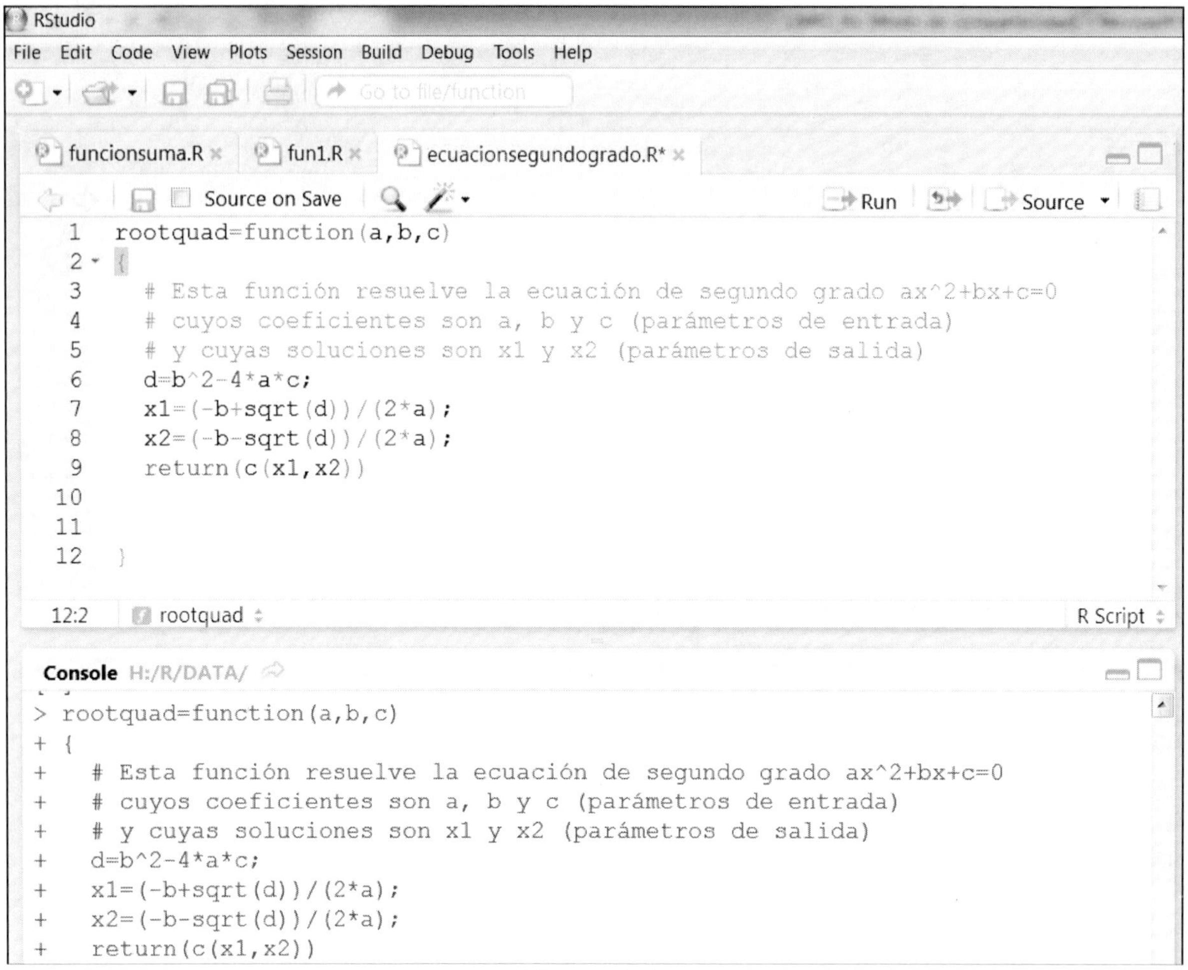

Figure 4-6

Now we will solve the equation *x^2-6x+2=0 in the* following way:

```
> rootcuad(1,-6,2)
[1] 5.6457513 0.3542487
```

There are several R commands that allow access to the different elements of the structure of a function. The most used are the following:

formals	*Allows access to the arguments of a function*
alist	*Allows the modification of the arguments of a function*
body	*Allows access to the body of a function*
edit	*Allows you to modify functions*
mapply(f,vec1,..,vecn	*Applies the function f to the specified vectors*

As a first example, let's look at the arguments of the *rootquad* function defined above

```
> formals(rootquad)
$a


$b


$c
```

Now we will access the body of the function.

```
> body(rootquad)
{
     d = b^2 - 4 * a * c
     x1 = (-b + sqrt(d))/(2 * a)
     x2 = (-b - sqrt(d))/(2 * a)
 return(c(x1, x2))
}
```

To modify any part of the function we use *edit*:

```
> edit(rootquad)
```

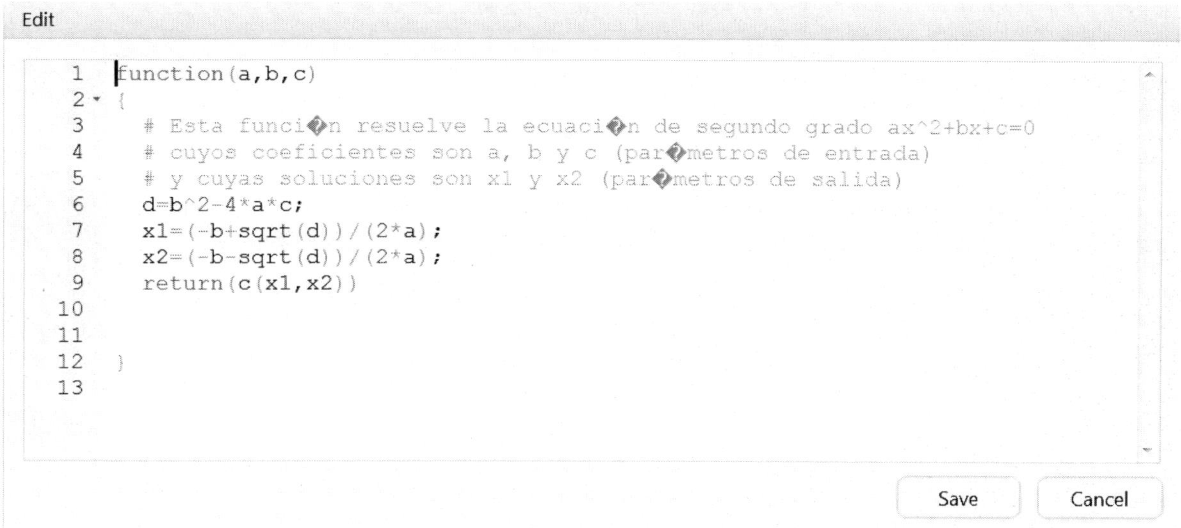

We can use *mapply* to solve the equation *x2+3x+2=0.*

```
> mapply(rootquad,1,3,2)
     [,1]
[1,]   -1
[2,]   -2

> rootquad(1,3,2)
[1] -1 -2
```

4.4 PREDEFINED MATH FUNCTIONS IN R

Apart from the general R commands, which are sometimes also called functions, the program implements various types of strictly mathematical functions. Some of them were already studied in previous chapters and the remaining ones will be dealt with in this and later chapters. To ask the program for help in using any of the functions, just type *"? function_name"* in the R command window. You can also use the program's general help (*Help* option in the general menu of the command window).

Among the mathematical functions of Matlab, the specialized functions of real variable and the statistical functions studied in the following paragraphs stand out:

4.4.1 Elementary mathematical functions of real variable

The group of elementary functions of the basic module of R presented in the following table were already studied in the chapter on the definition of variables (both scalar, vector and matrix). However, only as a summary, they are presented again here.

abs	*Module or absolute value*
sin, sinh	*Hyperbolic breast and sinus*
cos, cosh	*Hyperbolic cosine and cosine*
tan, tanh	*Tangent and hyperbolic tangent*
asin, asinh	*Sine Arch and Hyperbolic Sine Arch*
acos, acosh	*Cosine arc and hyperbolic cosine arc*
atan, atanh	*Arc tangent and hyperbolic arc tangent*
atan2	*Arc tangent in the fourth quadrant*
ceiling	*Rounding to the previous integer*
floor	*Rounding to the next integer (whole part)*
round	*Rounding to the nearest whole*
round(*x*, digits=n)	*Rounding to n digits*
signif(*x*, digits=n)	*Rounding to n significant digits*
exp	*Exponential*
expm1(x)	*exp(x)-1*
log	*Natural logarithm (Nepalese)*
log2	*Logarithm in base 2*
log10	*Decimal logarithm*
log1p(x)	*Logarithm of 1+x*
round	*Rounding to the nearest integer*
sign(x)	*Sign of x (1 if x is positive and 0 if x is negative)*
choose(a,b)	*Combined number a over b with whole a and b*
lchoose(a,b)	*Logarithm of the combination number a over b*

factorial(x)	*Factorial of integer x*
lfactorial(x)	*Logarithm of the factorial of x*
sqrt	*Square root*

As a first example we calculate the combinations without repetition of 10 elements taken from 4 and 4.

```
> choose(10,4)
[1] 210
```

The sine and cosine of the angles from 0 to 2π of /2π is then calculated by /2π.

```
> sin(0:pi/2:2*pi)
1] 0.000000e+00 1.000000e+00 1.224606e-16 -1.000000e+00

> round(sin(0:pi/2:2*pi))
[1] 0  1 0 -1

> round(cos(0:pi/2:2*pi))
[1] 1  0 -1 0
```

Some properties of the exponential and logarithmic functions are tested below.

```
> exp(log(2))
[1] 2

> log(exp(2))
[1] 2

> log2(2^1000)
[1] 1000

> log(7^70,7)
[1] 70
```

Some properties of trigonometric and hyperbolic functions are tested below.

```
> sin(pi)^2+cos(pi)^2
[1] 1
> (exp(5)+exp(-5))/2
[1] 74.20995

> cosh(5)
[1] 74.20995
```

```
> cosh(5)^2-sinh(5)^2
[1] 1


> 1+tan(pi/4)^2
[1] 2

> 1/sin(pi/4)^2
[1] 2
```

4.4.2 Basic statistical analysis functions with matrix argument

. The basic module of R has a group of functions related to data analysis in general, correlation, finite differences, etc. The following table presents this type of functions.

sum(A)	Sum of the elements of vector A
product(A)	Product of the elements of vector A
max(A)	Maximum value of the elements of vector A
min(A)	Minimum value of the elements of vector A
pmax(A,B,C..)	Vector of maximum values of A,B,C. (max(A[i],B[i],C[i],...)
pmin(A,B,C...)	Vector of minimum values of A,B,C. (min(A[i],B[i],C[i],...)
cumprod(A)	Vector A Accumulated Product Vector
cumsum(A)	Vector of accumulated sums of the elements of vector A
cummax(A)	Vector of accumulated maxima of the elements of vector A
cummin(A)	Vector of accumulated minima of the elements of vector A
range(A)	Minimum and maximum values of the elements of vector A range(A) = c(min(A), max(A))
rank(A)	Vector with the ranges of the elements of vector A (place occupied by each element in A)
mean(A)	Arithmetic mean of the columns of matrix A
weighted. mean(A,W)	Weighted arithmetic mean of the elements of A with weighting vector W
median(A)	median of the elements of vector A
quantile(A,probs=0.k)	k-th quantification of the elements of vector A

IQR(A)	Interquartile range of the elements of vector A
fivenum(A)	Minimum, first quartile, median, third quartile and maximum of the elements of vector A
sd(A)	quasi-standard deviation of the elements of vector A
var(A)	Quasi-variance of the elements of vector A
cov(A)	Quasi-variance of the elements of vector A
var(A,B)	Covariance between Vectors A and B
cov(A,B)	Covariance between Vectors A and B
cor(A,B)	Correlation Coefficient between Vectors A and B
var(M)	Covariance matrix for the columns of the M
cov(M)	Covariance matrix for the columns of the M
cor(M)	Correlation matrix for the columns of the M
diff(A)	Generates the first difference vector from vector A
diff(A,lag=k)	Generates the vector k-that is the difference from vector A
scale(*x*, center=TRUE, scale=TRUE)	Standardizes the vector or matrix x

As an example, we start from the variable formed by the first 9 positive integers and calculate the variables of accumulated sums, accumulated products, accumulated maximums, accumulated minimums, first and second differences, maximum and minimum, ranges and the standardized variable.

```
> A=c(1, 2, 3, 4, 5, 6, 7, 8, 9)
> A
[1] 1 2 3 4 5 6 7 8 9
> cumprod(A)
[1] 1     2     6    24     120     720     5040 40320 362880
> cumsum(A)
[1] 1   3 6 10 15 21 28 36 45
> cummax(A)
[1] 1 2 3 4 5 6 7 8 9
> cummin(A)
[1] 1 1 1 1 1 1 1 1 1
> diff(A)
[1] 1 1 1 1 1 1 1 1
> diff(A,lag=2)
[1] 2 2 2 2 2 2 2
> range(A)
[1] 1 9
> rank(A)
[1] 1 2 3 4 5 6 7 8 9
> scale(A, center=TRUE, scale=TRUE)
```

```
              [,1]
 [1,] -1.4605935
 [2,] -1.0954451
 [3,] -0.7302967
 [4,] -0.3651484
 [5,]  0.0000000
 [6,]  0.3651484
 [7,]  0.7302967
 [8,]  1.0954451
 [9,]  1.4605935
attr(,"scaled:center")
[1] 5
attr(,"scaled:scale")
[1] 2.738613
```

We will now calculate the product, the sum, the maximum, the minimum , the mean, the median, the quasi-variance, the quasi-standard deviation, the 30th pe rcentile, the percentiles of orders 1, 25, 50, 75 and 100 and the interquartile range o f the variable A.

```
> product(A)
[1] 362880
> sum(A)
[1] 45
> max(A)
[1] 9
> min(A)
[1] 1
> mean(A)
[1] 5
> median(A)
[1] 5
> var(A)
[1] 7.5
> sd(A)
[1] 2.738613

> quantile(A,probs=0.3)
30%
3.4
> fivenum(A)
[1] 1 3 5 7 9
> IQR(A)
[1] 4
```

We then calculate the variance and standard deviation of the variable A.

```
> Variance=(length(A)-1)/length(A)*var(A)
> Variance
[1] 6.666667

> Typical deviation=sqrt((length(A)-1)/length(A)*var(A))
> Deviation from the norm
[1] 2.581989
```

Next, we calculate the correlation coefficient and the covariance between variables A and Sen(A).

```
> cor(A,sin(A))
[1] 0.001990017

> cov(A,sin(A))
[1] 0.00398658
```

Finally we consider the square matrix M of order 3 whose columns are three variables with consecutive positive integers. We will calculate the covariance matrix and the correlation matrix of the three variables.

```
> M=matrix(c(1,2,3,4,5,6,7,8,9),3,3)

> M
     [,1] [,2] [,3]
[1,]    1    4    7
[2,]    2    5    8
[3,]    3    6    9


> cov(M)
 [,1] [,2] [,3]
[1,]    1    1 1
[2,]    1    1    1
[3,]    1    1    1
> cor(M)
     [,1] [,2] [,3]
[1,]    1    1 1
[2,]    1    1    1
[3,]    1    1    1
```

Exercise 1. Build two functions as they allow finding the mean and variance of any variable

A very simple syntax to program the average would be the following:

To find the average of the first 9 natural numbers not null, the following command would be entered:

```
> average1(c(1,2,3,4,5,6,7,8,9))
[1] 5
```

A very simple syntax to program the variance would be the following:

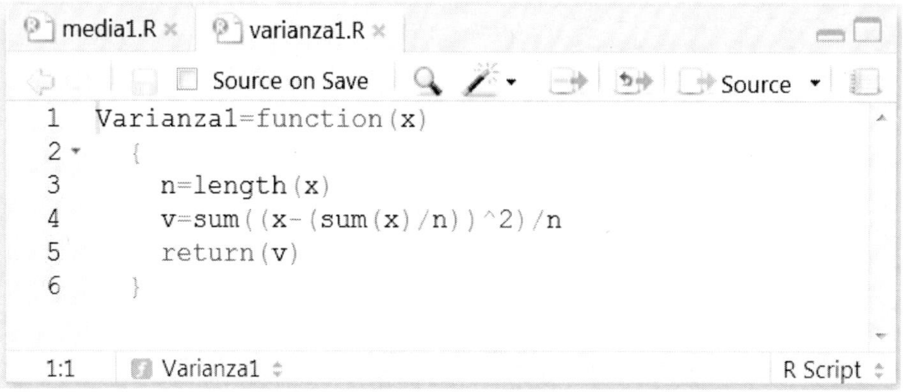

To find the variance of the first 9 natural numbers not null, the following command would be entered:

```
> Varianza1(c(1,2,3,4,5,6,7,8,9))
[1] 6.666667
```

Exercise 2. To build a function as they allow to find the covariance between two variables saving the problem of the missing values in the variables.

The syntax could be the following:

```
Covarianza<-function(x,y)
  {
    x=x[!is.na(x)]
    y=y[!is.na(y)]
    n=length(x)
    cv<-sum((x-(sum(x)/n))^2*(y-(sum(y)/n))^2)/n
    return(cv)
    }
```

We would calculate the covariance of two variables as follows:

```
> Covarianza(c(1,2,3,4),c(14,12,11,10))
[1] 4.609375
```

Exercise 3. The number of petals of 13 flowers of a given species is as follows: 8, 10, 6, 5, 8, 11, 8, 10, 7, 10, 7, 10 and 9. Calculate the mean, the variance and the coefficient of variation

We define the vector V whose components are the numbers of petals and then calculate the mean variance and coefficient of variation with the appropriate functions of R.

```
> V=c(8, 10, 6, 5, 8, 11, 8, 10, 7, 10, 7, 10, 9)

> mediaV=mean(V)
> mediaV
[1] 8.384615

> varianceV=(length(V)-1)/length(V)*var(V)
> varianceV
[1] 3.005917

> coefficient of variationV=sqrt(varianceV)/average(V)
> coefficient of variationV
[1] 0.2067785
```

The coefficient of variation has been calculated as the quotient between the standard deviation and the average:

Given the NP variable whose data are 8, 10, 6, 5, 8, 11, 8, 10, 7, 10, 7, 10 and 9, obtain their successive differences of orders 1 and 2.

We start by defining an NP vector containing the given values and then we use the MATLAB *diff* function properly.

```
> NP=c(8, 10, 6, 5, 8, 11, 8, 10, 7, 10, 7, 10, 9)

> DS1=diff(NP)
> DS1
 [1] 2 -4 -1 3  3 -3 2 -3 3 -3 3 -1

> DS2=diff(NP,lag=2)
> DS2
 [1] -2 -5 2  6 0 -1 -1 0  0 0  2
```

Exercise 5. An insurance agent sells policies to 5 individuals, all of the same age. According to the actuarial tables, the probability that an individual with that age will live 30 years longer is 3/5. Determine the probability that within 30 years they will live:

a) at least 3 individuals.
b) at most 2.

Since in 30 years time the situation of each individual will be whether he lives or not, and at least one of the two must be present, the situation for each individual is adjusted to a Bernoulli variable with probability of success (living 30 years more) equal to 3/5 = 0.6. When considering the 5 individuals, we are before a random variable X Binomial with $n = 5$ and $p = 0.6$, $X = B(5, 0.6)$. If we call $F(x)$ *to* the distribution function of $X = B(5, 0.6)$ at point x, the sections of the problem are calculated as follows:

a) We will have to calculate $P(X \geq 3)$, or what is the same, 1 -$p(X < 3)$ = 1-F(2).
b) We will have to calculate $P(X \leq 2)$, or what is the same $F(2)$.

The first probability is equivalent to:

$$1 - F(2) = 1 - \binom{5}{2} 0{,}6^2 0{,}4^3 - \binom{5}{1} 0{,}6^1 0{,}4^4 - \binom{5}{0} 0{,}4^5$$

which can be calculated by R as follows:

```
> 1-choose(5,2)*0.6^2*0.4^3-choose(5,1)*0.6*0.4^4-choose(5,0)
*0.4^5
[1] 0.68256
```

The second probability is equivalent to:

$$F(2)=\binom{5}{2}0{,}6^2 0{,}4^3+\binom{5}{1}0{,}6^1 0{,}4^4+\binom{5}{0}0{,}4^5$$

which can be calculated by R as follows:

```
> choose(5,2)*0.6^2*0.4^3+choose(5,1)*0.6*0.4^4+choose(5,0)*0.4
^5
[1] 0.31744
```

Exercise 6. The average number of cars arriving at a gasoline station is 210 per hour. If the station can handle a maximum of 10 cars per minute, determine the probability that more cars than it can handle will arrive at the fueling station in a given minute.

The random number of cars arriving at the service station in a minute can be represented by a Poisson variable *X of* parameter $m = 210/60 = 3.5$, since *m is* the average number of arrivals per minute (we had 210 arrivals per hour).

The probability that we are going to obtain will be given by *P*(*X*> 10), since in order for more cars per minute to arrive at the station than can be served, it is necessary for more than 10 cars per minute to arrive. But *P*(*X*> 10) = 1 - P(X 10≤) = 1 - F(*10*) being F *the* distribution function of a random Poisson variable of parameter 3.5.

To calculate the requested probability we will take into account that

$$1-F(10)=1-\sum_{k=0}^{10}\frac{e^{-3{,}5}3{,}5^k}{k!}$$

which can be calculated in R as follows:

```
> 1-exp(-3.5)*(1+3.5+3.5^2/factorial(2)+3.5^3/factorial(3)+ 3
.5^4/factorial(4)+3.5^5/factorial(5)+3.5^6/factorial(6)+ 3.5^
7/factorial(7)+3.5^8/factorial(8)+3.5^9/factorial(9)+ 3.5^10/
factorial(10))

[1] 0.001019394
```

Exercise 7. The proportion of individuals from a population with an income over two million pesetas is 0.005%. Determine the probability that among 5,000 individuals consulted there are at most 2 with that level of income, assuming that all those consulted respond.

Theoretically, the variable X defined by the number of individuals, among the 5,000, who have an income level higher than two million pesetas is a binomial variable with n = 5,000 and p = 0.00005. The requested probability would be $P(X$ $2\leq)$. But this calculation with such extreme numbers is annoying.

Now, given that p is very small (p = 0.00005) and that n is very large (n = 5,000), and also m = np = 0.25 < 5 and p < 0.1, we can already approximate the variable X = B(5,000, 0.00005) by a Poisson variable of parameter m = np = 5,000* 0.00005 = 0.25. Then P(X 2≤) we can calculate it by the Poisson distribution of parameter 0.25 as F(2), that is:

$$F(2) = \sum_{k=0}^{2} \frac{e^{-0,25} 0,25^k}{k!}$$

which can be calculated in R as follows:

```
> exp(-0.25)*(1+0.25+0.25^2/factorial(2))
[1] 0.9978385
```

Exercise 9. How many different triangles can be formed from the vertices of a dodecagon?

It is a matter of calculating the number of combinations of 12 elements taken 3 at a time. The following syntax is used for this purpose:

```
> a=choose(12,3)
> a
[1] 220
```

Exercise 10. Test for n=100 and p=30 that: $\binom{n}{p} + \binom{n}{p+1} = \binom{n+1}{p+1}$

The problem can be solved by checking the equality of both members using the following syntax:

```
> a=choose(100,30)+choose(100,30+1)
> a
[1] 9.569698e+25

> b=choose(100+1,30+1)
> b
[1] 9.569698e+25
```

Exercise 11. Be the variables X, Y and Z whose values are the following:

X	Y	Z
2	4	2
3	5	4
6	10	6
8	11	7
10	15	10

a) Calculate the mean, median, standard deviation, variance, and coefficient of variation for the three variables, and find a confidence interval for the mean based on each variable with a 95% confidence level
b) Determine the correlation matrix, deducing from it the degree of dependence between the variables
c) Determine the covariance matrix and interpret the result

We begin by introducing the variables X, Y and Z as columns of the matrix A.

```
> A=matrix(c(2,3,6,8,10,4,5,10,11,15,2,4,6,7,10),5,3)
> A
     [,1] [,2] [,3]
[1,]    2    4    2
[2,]    3    5    4
[3,]    6   10    6
[4,]    8   11    7
[5,]   10   15   10
```

To calculate the requested statistics we use the following syntax R:

```
> mean(A)
[1] 6.866667
> mean(A[,1])
[1] 5.8
> mean(A[,2])
[1] 9
> mean(A[,3])
[1] 5.8

> median(A[,1])
[1] 6
> median(A[,2])
[1] 10
> median(A[,3])
[1] 6

> varianceA1=4/5*var(A[,1])
> varianceA1
```

```
[1] 8.96
> varianceA2=4/5*var(A[,2])
> varianceA2
[1] 16.4
> varianceA3=4/5*var(A[,3])
> varianceA3
[1] 7.36

> devtipA1=sqrt(varianceA1)
> desvtipA1
[1] 2.993326
> devtipA2=sqrt(varianceA2)
> desvtipA2
[1] 4.049691
> devtipA3=sqrt(varianceA3)
> desvtipA3
[1] 2.712932

> cvA1=desvtipA1/mean(A[,1])
> cvA1
[1] 0.5160907
> cvA2=desvtipA2/mean(A[,2])
> cvA2
[1] 0.4499657
> cvA3=desvtipA3/mean(A[,3])
> cvA3
[1] 0.4677469
```

To calculate the confidence intervals we use the following syntax:

```
> c(mean(A[,1])-2*desvtipA1,mean(A[,1])+2*desvtipA1)
[1] -0.1866518 11.7866518

> c(mean(A[,2])-2*desvtipA2,mean(A[,2])+2*desvtipA2)
[1] 0.9006173 17.0993827

> c(mean(A[,3])-2*desvtipA3,mean(A[,3])+2*desvtipA3)
[1] 0.374136 11.225864
```

The intervals will be: [-0,187, 117867], [0,9006, 17,0994] y [0,3741, 11,2259]

To find the correlation matrix we use the following syntax R:

```
> cor(A)
          [,1]      [,2]      [,3]
[1,] 1.0000000 0.9899319 0.9802123
[2,] 0.9899319 1.0000000 0.9830213
[3,] 0.9802123 0.9830213 1.0000000
```

In the previous output it is observed that the correlation coefficient between *X* and *Y is* 0.9899, between *X* and *Z* 0.9802 and between Y and Z 0.9830, which indicates the strong positive linear dependence between each pair of variables.

To find the covariance matrix we use the following syntax R:

```
> cov(A)
      [,1] [,2] [,3]
[1,] 11.20 15.0  9.95
[2,] 15.00 20.5 13.50
[3,]  9.95 13.5  9.20
```

In view of the results we can say that as all covariances are positive, the pairs of variables vary in the same direction around their means.

Exercise 4-14. For a given country, be the variables R=Personal income in monetary units and H=Number of people going on vacation abroad and whose values are as follows:

R	*H*
1000	30
1100	50
1300	60
1500	70
1800	90
2000	100
2200	110
2500	120

Justify if a linear statistical dependence between R and H can be accepted.

We begin by introducing the variables R and H as column vectors, and then the correlation coefficient is calculated as follows:

```
> R=c(1000,1100,1300,1500,1800,2000,2200,2500)
> H=c(30,50,60,70,90,100,110,120)
> cor(R,H)
[1] 0.986152
```

It is observed that the correlation coefficient between R and H is r = 0.9862, which ensures the strong positive linear dependence between the two variables.

Chapter 5

STATISTICAL GRAPHS FOR EXPLORATORY DATA ANALYSIS

5.1 EXPLORATORY DATA ANALYSIS

Before applying any data analysis technique, it is necessary to perform a previous analysis of the information available. It is necessary to examine individual variables and the relationships between them, as well as to evaluate and solve problems in the design of the research and in the collection of data. The first task that is usually addressed is the *exploratory and graphic analysis of the data*. Most statistical *software* has tools that provide ready-made graphical techniques for data examination that are enhanced with more detailed statistical measurements for description. These techniques allow the examination of the distribution characteristics of the variables involved in the analysis, the bivariate (and multivariate) relationships between them and the analysis of differences between groups. It should be kept in mind that graphic representations never replace formal statistical diagnostic measures (data adjustment contrasts to a distribution, asymmetry contrasts, randomness contrasts, etc.), but provide an alternative way to develop a perspective on the character of the data and the interrelations that exist, even if they are multivariate.

The techniques of exploratory data analysis allow us to analyze the information exhaustively and to detect the possible anomalies that the observations present. J. W. Tuckey has been one of the pioneers in introducing this type of analysis. The most commonly used descriptive statistics have been the mean and standard deviation. However, the automatic use of these indices is not very advisable. The mean and standard deviation are convenient indexes only when the distribution of data is approximately normal or at least symmetric and unimodal. But the variables under study do not always meet these requirements. Therefore a thorough examination of the data structure is necessary.

It is recommended to start an exploratory data analysis with graphics that allow to visualize its structure. We are in front of the ***visual exploration*** tools. However, for ***formal exploration,*** the use of *robust* (or resistant) *statistics* is highly advisable when the data do not fit a normal distribution. These statistics are those that are little affected by outliers. They are usually based on the median and the quartiles and are easy to calculate. As a result of the exploratory analysis, it is sometimes necessary to carry out transformation of variables.

For quantitative data it is advisable to start with the stem and leaf chart or digital histogram. The next step is usually to examine the possible presence of normality, symmetry and outliers in the data set. For this purpose, box and mustache graphs are usually used. However, box plots should always be accompanied by digital histograms (or stem and leaf plots), since the former do not detect the presence of multimodal distributions. Scatter plots give us an idea of the relationships between variables and their adjustment.

5.1.1 Frequency Histogram

Anyway, it is always convenient to start the exploratory data analysis with the construction of the associated frequency histogram, in order to be able to intuit the probability distribution of the data, its normality, its symmetry and other interesting properties in the data analysis. As an example we can consider the variable X defined as the fuel consumption in liters at 1000 kilometers of the cars of a given brand. The values for X are as follows:

43,1	36,1	32,8	39,4	36,1	19,9	19,4	20,2	19,2	20,5	20,2	25,1	20,5	19,4	20,6
20,8	18,6	18,1	19,2	17,7	18,1	17,5	30	27,5	27,2	30,9	21,1	23,2	23,8	23,9
20,3	17	21,6	16,2	31,5	29,5	21,5	19,8	22,3	20,2	20,6	17	17,6	16,5	18,2
16,9	15,5	19,2	18,5	31,9	34,1	35,7	27,4	25,4	23	27,2	23,9	34,2	34,5	31,8
37,3	28,4	28,8	26,8	33,5	41,5	38,1	32,1	37,2	28	26,4	24,3	19,1	34,3	29,8
31,3	37	32,2	46,6	27,9	40,8	44,3	43,4	36,4	30,4	44,6	40,9	33,8	29,8	32,7
23,7	35	23,6	32,4	27,2	26,6	25,8	23,5	30	39,1	39	35,1	32,3	37	37,7
34,1	34,7	34,4	29,9	33	34,5	33,7	32,4	32,9	31,6	28,1	30,7	25,4	24,2	22,4
26,6	20,2	17,6	28	27	34	31	29	27	24	23	36	37	31	38
36	36	36	34	38	32	38	25	38	26	22	32	36	27	27
44	32	28	31											

To explore this information we elaborated the frequency table associated to the data and studied the possible normality and symmetry of the fuel consumption distribution. As it is a quantitative variable with 154 values between 13 and 49, it will be necessary to group them in intervals or classes. To do this we take 12 intervals of equal width (12 is an integer that approximates well the square root of N = 154). The width of the intervals will be (49 - 13)/12 = 3. The frequency table in Figure 5-1 is obtained.

Interval	Limit below	Limit top	Brand Classroom	nor	$fi = ni/N$	Nor	$Fi = ni/N$
1	13,0	16,0	14,5	1	0,0065	1	0,0065
2	16,0	19,0	17,5	14	0,0909	15	0,0974
3	19,0	22,0	20,5	22	0,1429	37	0,2403
4	22,0	25,0	23,5	15	0,0974	52	0,3377
5	25,0	28,0	26,5	22	0,1429	74	0,4805
6	28,0	31,0	29,5	16	0,1039	90	0,5844
7	31,0	34,0	32,5	22	0,1429	112	0,7273
8	34,0	37,0	35,5	22	0,1429	134	0,8701
9	37,0	40,0	38,5	11	0,0714	145	0,9416
10	40,0	43,0	41,5	3	0,0195	148	0,9610
11	43,0	46,0	44,5	5	0,0325	153	0,9935
12	46,0	49,0	47,5	1	0,0065	154	1,0000

Figure 5-1

We have observed the 154 values on the consumption of cars that initially did not provide much information. Obviously there is variability in car consumption; however, it is very difficult to detect what pattern this variability follows in order to better determine the structure of the data. Therefore, first of all it has been convenient to carry out an arrangement of the data according to its magnitude, that is to say, a table of frequencies, which will provide some light on the underlying frequency distribution.

The next task is the construction of the frequency histogram, a graph suitable for a quantitative variable with its values grouped in intervals. Its representation is presented in Figure 5-2.

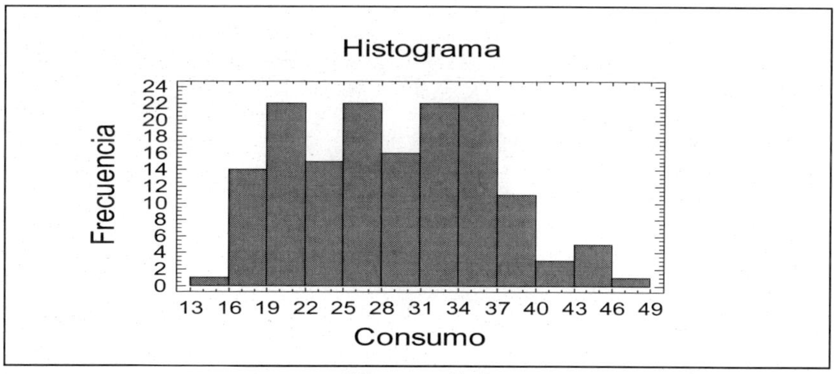

Figure 5-2

It is observed that the underlying distribution that models the data on the variable consumption of cars is approximately symmetrical and adjustable to a bell shape, which allows us to think about the existence of normality and symmetry in the distribution of *X*. We see that the histogram gives a clear idea of the distribution of the variable, including a probabilistic model for its modeling, in this case the normal distribution. The simple examination of the tabulated data initially did not provide any information, but its graphing gives light to the process.

5.1.2 Stem and leaf diagram

The stem-leaf diagram is a semi-graphic procedure for presenting information for quantitative variables, which is especially useful when the total number of data is small (less than 50). The principles for making the diagram (due to Tukey) are as follows:

- Round off the data to two or three significant figures.
- Arrange them in two columns separated by a vertical line so that for data with two digits the tens figure is on the left of the vertical line (stem of the diagram), and on the right the units (leaves or branches of the diagram). For example, 87 will be written 8 7. For three digit data the stem will be formed by the hundreds and tens digits, which will be written on the left of the vertical line, and the leaves will be formed by the units digit, which will be written on the right of the vertical line.
- Each stem defines a class, and is written only once. To the right of it, the successive sheets corresponding to that stem are written in order. The number of leaves for each stem represents the frequency of each class.

The stem-leaf diagram, also called *digital histogram,* is a combination of a bar chart and a frequency table. By maintaining the variable values, the stem-leaf diagram is more informative than the classic bar histogram, since it preserves the original data and, at the same time, composes a profile that helps to study the shape and symmetry of the distribution. It is therefore an exploratory data analysis tool that shows the range of the data, where they are most concentrated, their symmetry and the presence of outliers. This procedure is not very advisable for large data sets.

Below is the stem and leaf diagram (Figure 5-3) for the variable *X* related to the variable consumption of the cars defined in the previous section.

```
Diagrama de Tallo y Hojas para X: unidad = 1,0    1|2 representa 12,0

   23      1|56667777777888888999999
   51      2|0000000000111222333333333444
  (31)     2|5555566666777777777788888899999
   72      3|00000111111122222222222233334444444444
   34      3|555666666666777777888889999
    9      4|00133444
    1      4|6
```

Figure 5-3

The range of *X* has been divided into 7 classes or intervals called *stems,* each represented by a row in the diagram. The first number in each row (separated from the others) shows the absolute frequency of the corresponding class. The second number of each row presents the tens of each value of *X in* its corresponding class. The rest of the numbers in each row (called *leaves*) are the unit numbers of all elements of the class defined by the row. In this way, besides presenting the distribution of the elements in the form of a horizontal histogram, the diagram shows the elements themselves. The sheets allow the analysis of symmetry, normality and other characteristics of the distribution in the same way as a histogram.

5.1.3 Box and Whisker Chart

The box and whiskers graph allows you to analyze and summarize a given univariate data set. This exploratory data analysis tool will allow studying the symmetry of the data, detecting outliers and glimpsing an adjustment of the data to a given frequency distribution.

The box and whiskers graph divides the data into four areas of equal frequency, a central box divided into two areas by a vertical line and two other areas represented by two horizontal segments (whiskers) starting from the center of each vertical side of the box. The central box encloses 50 percent of the data. The system draws the median as a vertical line inside the box. If this line is in the center of the box there is no asymmetry in the variable. The vertical sides of the box are located in the lower and upper quartiles of the variable. Starting from the center of each vertical side of the box the two whiskers are drawn, one to the left and the other to the right. The whisker on the left has one end in the first quartile *Q1,* and the other in the value given by the first quartile minus 0.5 times the interquartile range, that is, $Q1 - 1.5*(Q3 - Q1)$.

The right whisker has one end in the third quartile *Q3* and the other end in the value given by the third quartile plus 1.5 times the interquartile range, i.e. $Q3+1.5*(Q3 - Q1)$. The system considers *outliers* those to the left of the left whisker and to the right of the right whisker. The system separates these data from the rest and represents them by means of points aligned with the central horizontal line so that they are easy to detect. Inside the central box the average is represented with a plus sign.

Below is the box and whiskers graph for the variable *X* related to the variable consumption of cars (Figure 5-4).

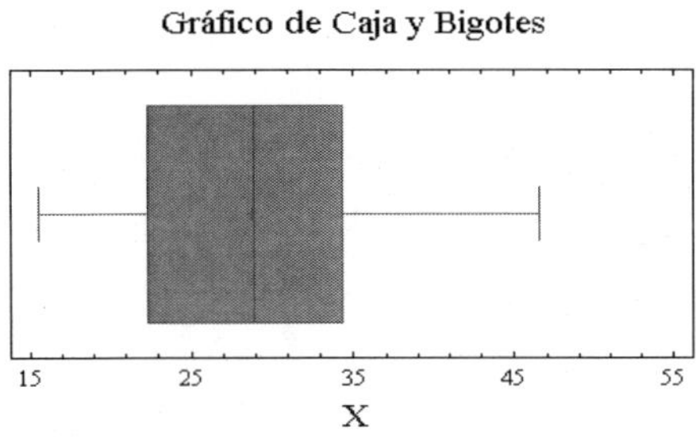

Figure 5-4

The graph shows that the variable *X* (consumption of cars every 1000 kilometers) varies between 15.5 and 46.6 and that the central 50% of cars consume between 22 (first quartile) and 34.5 (third quartile) liters at 1000 kilometers.

On the other hand, there are no abnormally large *X* values (*outliers),* since in the Figure no points appear aligned with the whiskers. The distribution is slightly asymmetrical to the right, since the area on the right in the central area of the Figure is larger than the one on the left. The median corresponds approximately to the value 29 of *X.*

The box and whiskers graphics allow the option of *median notch* representation, which places two notches on the ends of the median.

The width of the notch represents an approximate confidence interval for the median with a 95% confidence coefficient (α= 0.05) which is determined by the expression $M\pm$ *(1.*25R/1.35√ *n)* (1+1/√ 2)Z /2/2$_\alpha$, where *R is* the interquartile range of the variable, *M is the* median, and $Z_{/2\alpha}$is the value of the normal distribution (0.1) which leaves /2α of probability to the right.

For our variable *X,* car consumption, we will have a box and whiskers graph with notches in Figure 5-5.

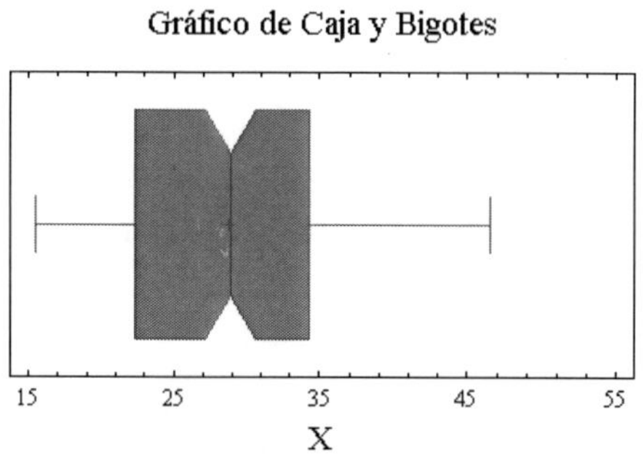

Figure 5-5

5.1.4 Multiple Box and Whisker Chart

In statistics it is typical to divide the data set of a variable into rational subgroups, which can be for example strata defined according to a certain stratification variable. The multiple box and whiskers graph will allow to analyze, summarize and compare simultaneously several given univariate data sets, corresponding to the different groups in which the values of a variable can be subdivided. This exploratory data analysis tool will allow studying data symmetry, detecting outliers and representing means, medians, ranges and extreme values for all groups. As the representation is simultaneous for all data sets, you will be able to compare means, medians, ranges, extreme values, symmetries and outliers for all groups. The multiple graph will represent horizontally a box and whiskers chart for each group of values of the variable under study.

If we classify the consumption of cars (variable X) according to their displacement (variable Y), we are making subgroups with the values of X according to the stratification variable Y. The possible values of Y are 8, 6, 5, 4 and 3 cylinders. The values of X for each value of Y are given below:

X	Y	X	Y	X	Y	X	Y	X	Y	X	Y	X	Y	X	Y	X	Y	X	Y
43,1	4	36,1	4	32,8	4	39,4	4	36,1	4	19,9	8	19,4	8	20,2	8	19,2	6	20,5	6
20,2	6	25,1	4	20,5	6	19,4	6	20,6	6	20,8	6	18,6	6	18,1	6	19,2	8	17,7	6
18,1	8	17,5	8	30	4	27,5	4	27,2	4	30,9	4	21,1	4	23,2	4	23,8	4	23,9	4
20,3	5	17	6	21,6	4	16,2	6	31,5	4	29,5	4	21,5	6	19,8	6	22,3	4	20,2	6
20,6	6	17	8	17,6	8	16,5	8	18,2	8	16,9	8	15,5	8	19,2	8	18,5	8	31,9	4
34,1	4	35,7	4	27,4	4	25,4	5	23	8	27,2	4	23,9	8	34,2	4	34,5	4	31,8	4
37,3	4	28,4	4	28,8	6	26,8	6	33,5	4	41,5	4	38,1	4	32,1	4	37,2	4	28	4
26,4	4	24,3	4	19,1	6	34,3	4	29,8	4	31,3	4	37	4	32,2	4	46,6	4	27,9	4
40,8	4	44,3	4	43,4	4	36,4	5	30,4	4	44,6	4	40,9	4	33,8	4	29,8	4	32,7	6
23,7	3	35	4	23,6	4	32,4	4	27,2	4	26,6	4	25,8	4	23,5	6	30	4	39,1	4
39	4	35,1	4	32,3	4	37	4	37,7	4	34,1	4	34,7	4	34,4	4	29,9	4	33	4
34,5	4	33,7	4	32,4	4	32,9	4	31,6	4	28,1	4	30,7	6	25,4	6	24,2	6	22,4	6
26,6	8	20,2	6	17,6	6	28	4	27	4	34	4	31	4	29	4	27	4	24	4
23	4	36	4	37	4	31	4	38	4	36	4	36	4	36	4	34	4	38	4
32	4	38	4	25	6	38	6	26	4	22	6	32	4	36	4	27	4	27	4
44	4	32	4	28	4	31	4												

The multiple box and whiskers graph of the consumption of cars (variable *X)* according to their displacement (variable *Y),* is shown in Figure 5-6.

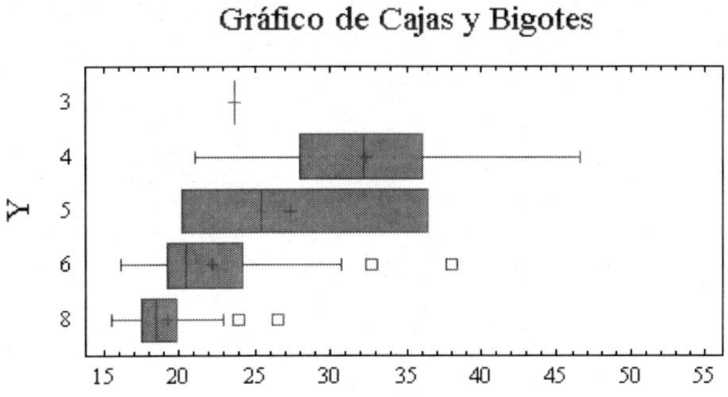

Figure 5-6

The graph allows us to say that the variable *X* (liters consumed at 1000 kilometers) *for 8-cylinder cars* varies between 15.5 and 23, and that the central 50% of these cars consume between 17.5 (first quartile) and 20 (third quartile) liters at 1000 kilometers, there are 2 abnormally large values of *X (outliers),* as in the figure there are two points aligned with the mustache on the right. The distribution of *X* for

8-cylinder cars is slightly asymmetrical to the right, since the area on the right in the central area of the Figure is larger than the left, and the median corresponds approximately to the value of 18.5 *X, with the* average being approximately 19.5.

For 6-cylinder cars, the liters consumed every 1000 kilometers (variable *X*) vary between 16 and 31, concentrating 50% central X values between 19 (first quartile) and 24 (third quartile), there are 2 *outliers of X* abnormally large (*outliers),* as in the figure there are two points aligned with the mustache on the right. The distribution of *X* for 6-cylinder cars is asymmetrical to the right, the median of *X is* close to 20.5 and the average to 22.5.

For 5-cylinder cars, the liters consumed every 1000 kilometers (variable *X*) vary between 20.2 and 36.5, with 50% of the *X* values concentrated in the middle between the same values, with no whiskers or *outliers*. The distribution of *X* for 5-cylinder cars is asymmetrical to the right, the median *X is* close to 25.5 and the average is 27.5.

For 4-cylinder cars, the liters consumed every 1000 kilometers (variable *X*) vary between 21 and 47, with the central 50% of *X* values concentrated between 28 (first quartile) and 36 (third quartile), with no *outliers*. The distribution of *X* for 4-cylinder cars is practically symmetrical with median and average values of approximately 32.

For 3-cylinder cars there is a single *X* value, which does not allow the box and whiskers graph to be constructed.

If we compare the different graphs, we see that the asymmetry of *X is* stronger for 5 and 6-cylinder cars, for 8-cylinder cars it is less and for 4-cylinder cars it does not exist. Abnormally large *X* values only appear for 6 and 8 cylinder cars. The averages and medians vary quite a bit for the different groups of *X values* determined by the *Y* values.

5.1.5 Symmetry and Normality Chart

The symmetry graph is a tool that allows you to visually analyze the degree of symmetry of a variable. On the abscissa axis are represented the distances of the values of the variable to the median that are below it, and the ordinate axis represents the distances of the values of the variable to the median that are above it. If the symmetry were perfect, the resulting set of points would be the main diagonal. The closer the graph is to the diagonal, the more symmetry there will be in the distribution of the variable.

For the example of the variable *X, a* variable defined by the number of liters consumed by cars every 1000 kilometers that we have been considering throughout the chapter, we have the Symmetry Graph in Figure 5-7.

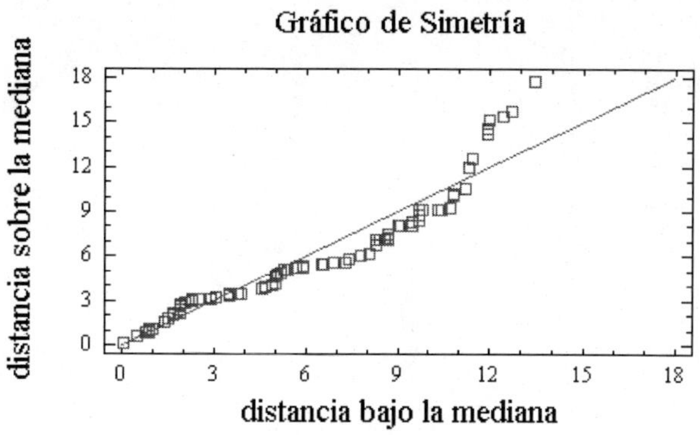

Figure 5-7

The graph made with the median can also be made with quantiles and percentiles (QQ and PP graphs respectively). The PP and QQ graphs are usually used to test the normality of the data. The variable will be more normal when the point cloud is better adjusted to the diagonal of the first quadrant.

5.1.6 Scatter plot

It is a graph that allows you to see the relationship between two or more variables. It is formed by points whose Cartesian coordinates are the pairs of values of two variables whose relationship we want to study represented one on the vertical axis and another on the horizontal axis.

The positioning of the points in the scatter plot defines the relationship between the variables. If they are placed around a line, there is a linear correlation between the variables. If the points follow a non-linear pattern, the relationship between the variables cannot be defined as linear. If the point cloud is random and scattered, there is no relationship between the variables. A scatter plot for a multivariate analysis that offers a lot of information is the one that represents in a matrix structure the scatter plots of all possible pairs of variables (lower triangular area of the matrix), their frequency histograms (diagonal of the matrix) and the correlation coefficients of all pairs of variables under study (upper triangular area of the matrix).

Figure 5-8 shows the matrix graph for the variables *Consumption*, *Power* and *Price of the* cars.

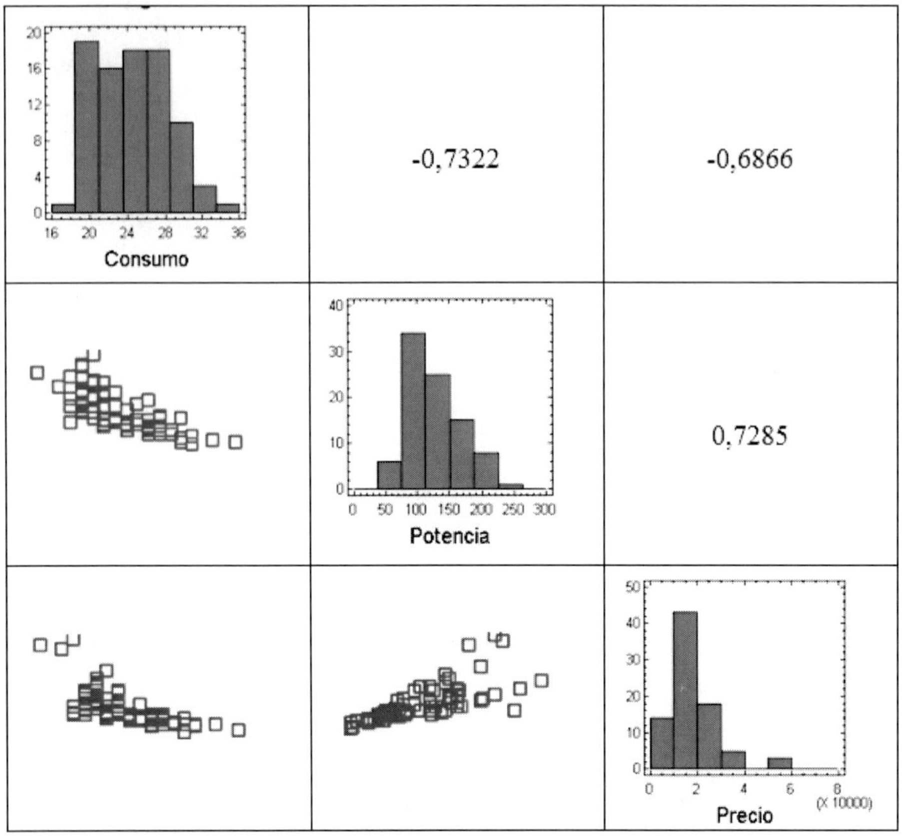

Figure 5-8

The consumption scatter plot with the power located in the element (2.1) of the matrix is easily adjustable to a negative slope line, indicating a strong negative relationship between both variables. This information is corroborated by the correlation coefficient between consumption and power (–0.7322) located in the element (1.2) of the matrix. The consumption scatter plot with the price located in the element (3.1) of the matrix is easily adjustable to a negative slope line, indicating a strong and negative relationship between both variables. This information is corroborated by the correlation coefficient between consumption and price (–0.6866) located in element (1.3) of the matrix. The power scatter plot with the price located in the element (3.2) of the matrix is easily adjustable to a positive slope line, indicating a strong and positive relationship between both variables. This information is corroborated by the correlation coefficient between power and price (0.7285) located in the element (2.3) of the matrix. The histograms of the main diagonal of the matrix indicate normality and symmetry for consumption and power, and some asymmetry for price.

5.1.7 Graphs for qualitative variables

The visual exploration of qualitative variables is usually carried out by means of rectangle diagrams, sector diagrams and pictograms.

The *rectangle diagrams* are constructed by assigning to each mode of the qualitative variable a rectangle with height equal to (or proportional to) its absolute frequency and with a constant base.

As an example, Figure 5-9 shows a rectangle diagram representing the assets according to the different modalities of the branch of activity variable. On each rectangle is presented the absolute frequency *or* in thousands of assets of the corresponding branch of activity. On the abscissa axis are presented the branches of activity themselves and on the ordinate axis are presented different values of absolute frequencies by intervals that serve as a reference to locate the height of each rectangle.

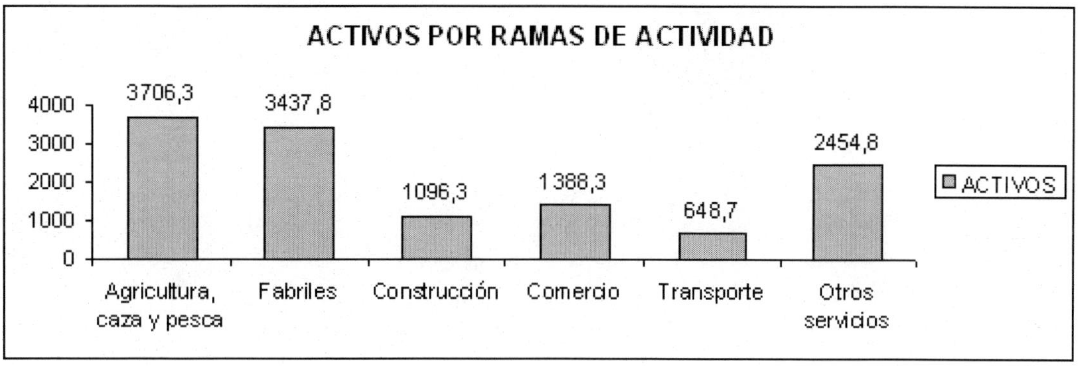

Figure 5-9

Sector (or pie) *diagrams* are the most commonly used type of graph to represent frequency distributions of qualitative variables. The variable is represented in a circle whose portions (circular sectors) have an area proportional to the absolute frequencies of the modalities of the variable. To make the graph of sectors it is sufficient to assign to each mode of the variable a circular sector whose central angle is proportional to the absolute frequency of the mode. Mathematically, the central angle α_i *of the i-th* mode whose absolute frequency is *nor* can be expressed as follows:

$$\alpha_i = kn_i \Rightarrow \frac{\alpha_i}{n_i} = k = \frac{\sum \alpha_i}{\sum n_i} = \frac{360}{N} \Rightarrow \alpha_i = \frac{360}{N} n_i = 360 \frac{n_i}{N} = 360 f_i$$

The calculations needed for our example are presented in the table in Figure 5-10.

BRANCH	ASSETS (n_i)	$f_i = n_i/N$	$\alpha_i = 360 f_i$
Agriculture, hunting and fishing	3706,3	0,29	104,79
Manufacturers	3437,8	0,27	97,20
Construction	1096,3	0,09	31,00

Commerce	1388,3	0,11	39,25
Transportation	648,7	0,05	18,34
Other services	2454,8	0,19	69,41
N=	12732,2		

Figure 5-10

Figure 5-11 presents the pie chart.

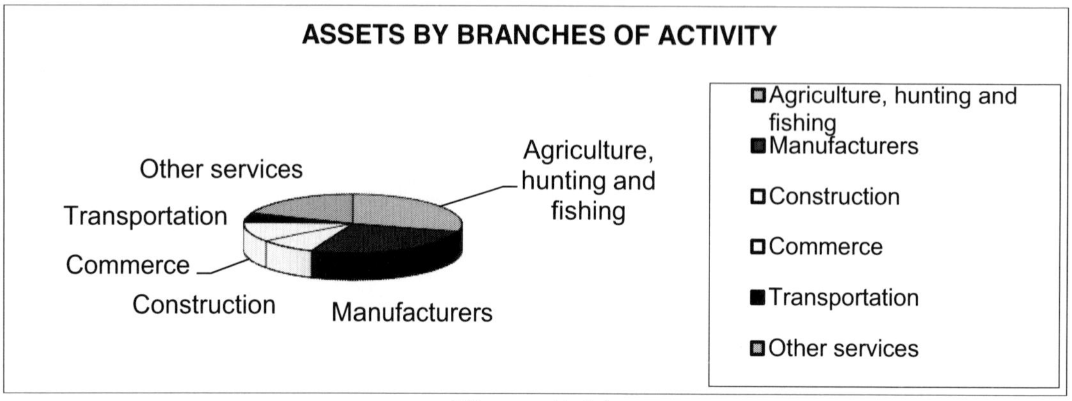

Figure 5-11

Another common way of constructing sector charts is to assign to the circular sector relative to the *i-th* mode a percentage equal to the percentage representing its absolute frequency *or* over the total frequency $N = \sum n_i$. Mathematically, the expression of the percentage p_i *relative* to the *i-th mode* is expressed as follows:

$$p_i = 100\frac{n_i}{N} = 100 f_i$$

Figure 5-12 shows the above pie chart with percentages.

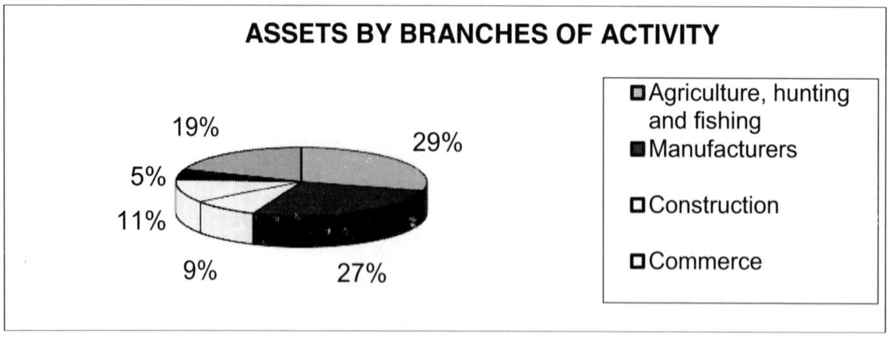

Figure 5-12

The *pictograms* are constructed representing in a pictorial way each modality of the qualitative variable indicating by a suggestive silhouette the meaning of each character unit. For example, Figure 5-13 shows a cylinder on each state on the map of the United States whose height is proportional to the number of facilities with hazardous error in 1997.

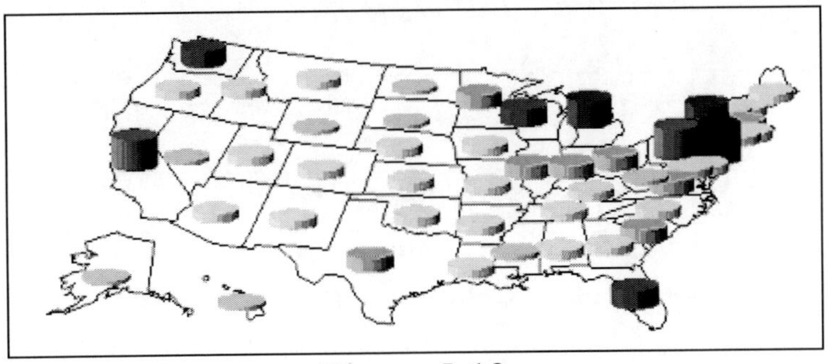

Figure 5-13

Any silhouette can be used in the pictograms to represent the frequency of each mode of the qualitative variable. Figure 5-14 shows the previous pictogram with a cone over each state whose height indicates the frequency of facilities with hazardous error.

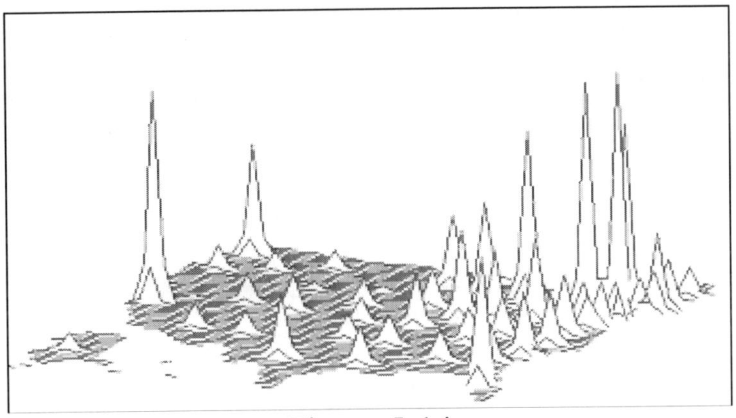

Figure 5-14

5.2 COMMANDS FOR EXPLORATORY GRAPHICS IN R. BASIC SYNTAX

The following is a list of R commands with their simplest syntax to perform exploratory data analysis

hist(x)	*It represents the histogram of the variable x*
stem(x)	*Stem and leaf chart of variable x*
barplot(x)	*Bar chart with heights given by the variable x*
dotchart(x)	*Point graph of the variable x*
plot(x, y)	*Scatter plot representing kla point cloud (xi, yi)*
foot(x, labels)	*Chart of sectors with areas of each sector given by the variable x y with the labels given for each area*

boxplot(x, y, z,...)	*It represents the box and whiskers graphs for the variables x, y, z, ...*
qqnorm(x)	*It represents the quantiles graph for the normality of the variable x*
qqplot(x,y)	*It represents the quantiles of y with respect to the quantiles of x*
qqline(x,y)	*Add expected lines to qqplot(x)*
plot(density(x))	*Graphs the kernel density function of the variable x*
pairs(X)	*Represents the bivariate scatter plots between the columns of the data frame X*
pairs(~x+y+z+...)	*Graph the bivariate scatter plots of x,y,z, ...*
plot(X)	*Represents the bivariate scatter plots between the columns of the data frame X*
plot(X[,1:k])	*Represents the bivariate scatter plots between the first k columns of the data frame X*
boxplot(X)	*Represents a box and whiskers graph for each variable in the data frame X*
coplot(y~x|f)	*Scatter plots of x and y for each value of the factor f*
boxplot(x~f)	*Make the box and whiskers graphs of the variable x for each level of the factor f*
boxplot(X,Y)	*Scatter plots of the first column of X against the first column of Y, of the second column of X against the second column of Y, etc.*
stripplot(x)	*Graph of x-values on one line (alternate to boxplot for small sample sizes)*
sunflowerplot(x,y)	*Similar to a plot, but the points with similar coordinates are drawn as flowers whose number of petals represent the number of points.*

To see some examples we consider the set of data *coches.sav in* SPSS format that is located in the subdirectory *H:Data of* the hard disk and that contains the variables related to the consumption (*consumption),* motorization (*engine),* power (*cv),* weight (*weight),* acceleration (*acel),* year of manufacture (*year),* geographic origin (*origin),* cylinder (*cylinder)* and a filter indicating failure (*derivative).*

First we read the file and enable its variables.

```
> library(foreign)
coches=read.spss("G:/R/data/coches.sav", use.value.labels = T
RUE)
> attach(cars)
```

To see the variables of the file and their properties we use the *summary* function.

```
> summary(cars)
          Length Class Mode
consumption 406     -none- numeric
engine      406     -none- numeric
cv         406     -none- numeric
weight      406     -none- numeric
acel       406     -none- numeric
year        406     -none- numeric
origin     406 numeric     factor
cylinder 406 numeric     factor
derivative 406 numeric     factor
```

To represent the consumption histogram (Figure 5-15) we use the following syntax:

```
> hist(consumption)
```

To represent the box and whiskers graph of consumption and acceleration (Figure 5-16) we use the following syntax:

```
> boxplot(consumption, acel)
```

To represent the consumption stem and leaves diagram (Figure 5-17) we use the following syntax:

```
> stem(consumption)
```

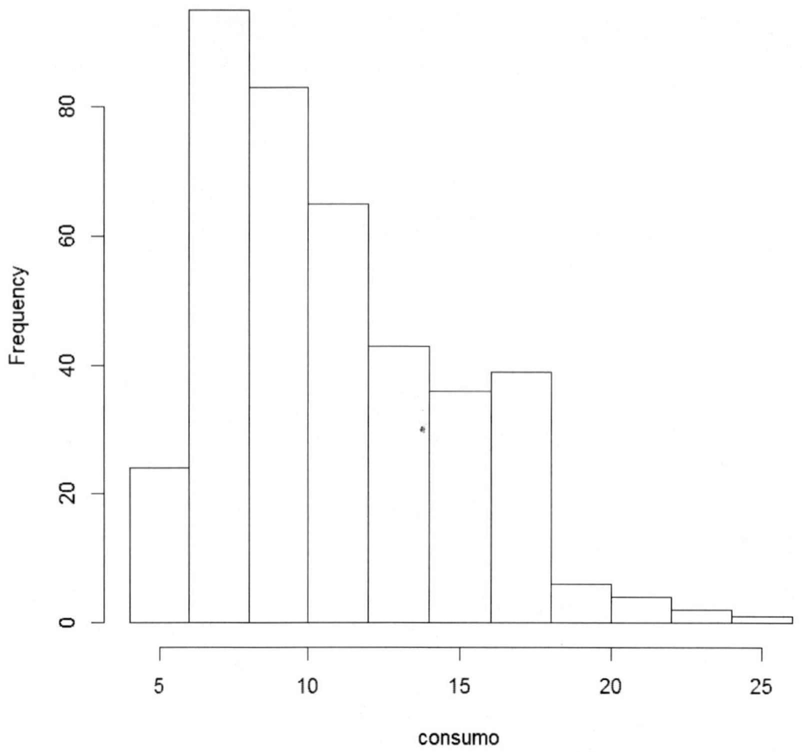

Figure 5-15

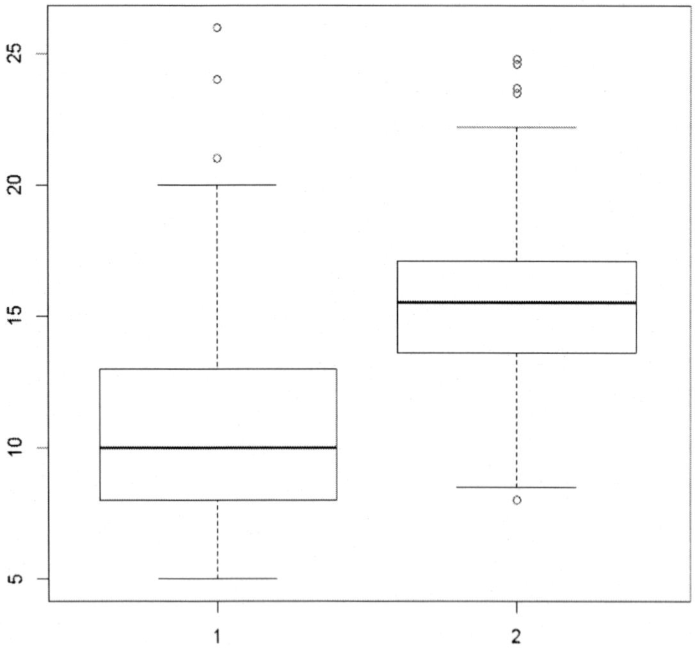

Figure 5-16

```
The decimal point is at the |

 5 | 000000
 6 | 000000000000000000
 7 | 0000000000000000000000000000000000000000000000000
 8 | 0000000000000000000000000000000000000000000000
 9 | 00000000000000000000000000000000000000000000000000
10 | 000000000000000000000000000000000
11 | 0000000000000000000000000000
12 | 0000000000000000000000000000000000000
13 | 00000000000000000000000000000000
14 | 00000000000
15 | 0000000000000000000
16 | 00000000000000000
17 | 0000000000000000000
18 | 00000000000000000000
19 |
20 | 000000
21 | 0000
22 |
23 |
24 | 00
25 |
26 | 0
```

Figure 5-17

To represent the scatter diagram of consumption and power (Figure 5-18) we use the following syntax

```
> plot(consumption,cv)
```

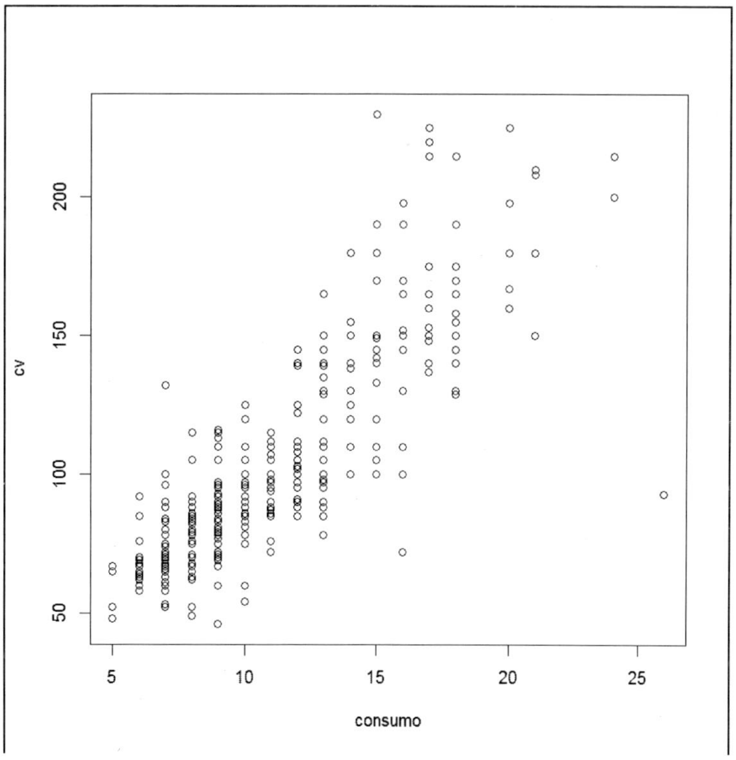

Figure 5-18

To represent the weight normality graph (Figure 5-19) we use the following syntax:

```
> qqnorm(weight)
```

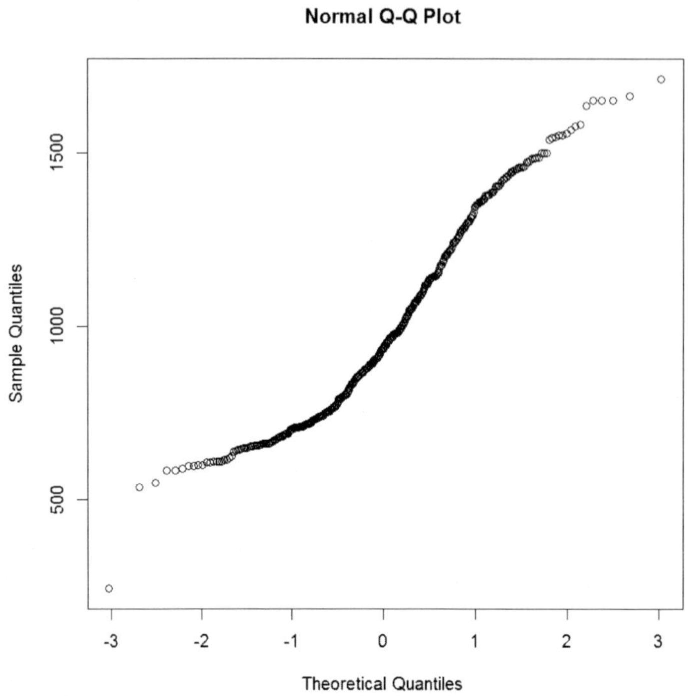

Figure 5-19

Next we will represent a multiple scatter plot for the variables consumption, power and weight.

First it is necessary to build a data frame with the three variables using the following syntax:

```
> data=data.frame(consumption,cv,weight)
```

The next task is to represent the multiple scatter plot (Figure 5-20) using any of the following three syntaxes:

```
> pairs(data)
> plot(data)
> plot(data[,1:3])
```

To represent a box and whiskers diagram for each variable of the data frame on the same graph (Figure 5-21), the following syntax is used:

```
> boxplot(data)
```

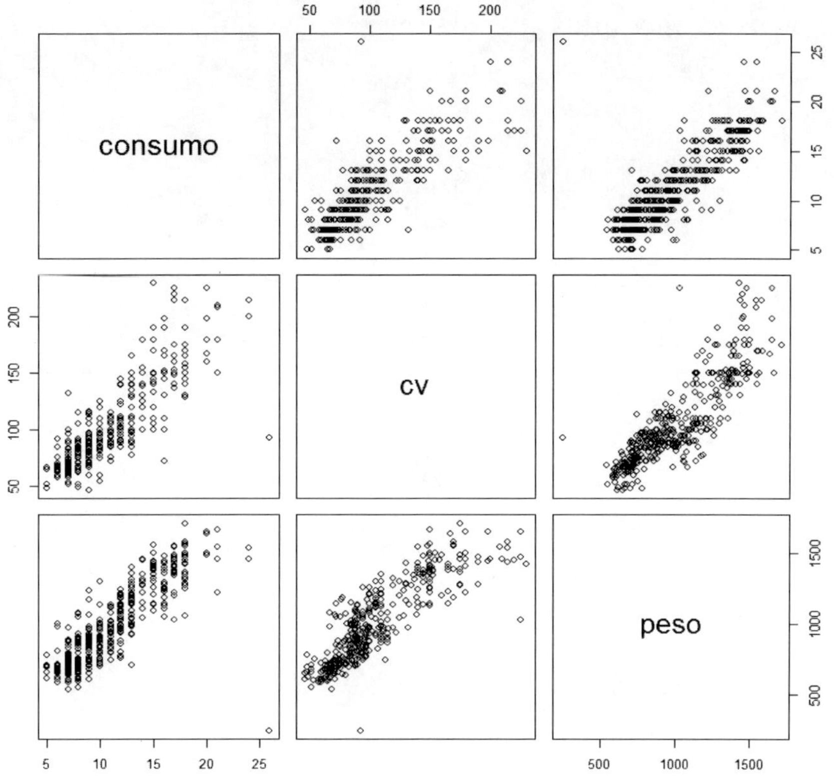

Figure 5-20

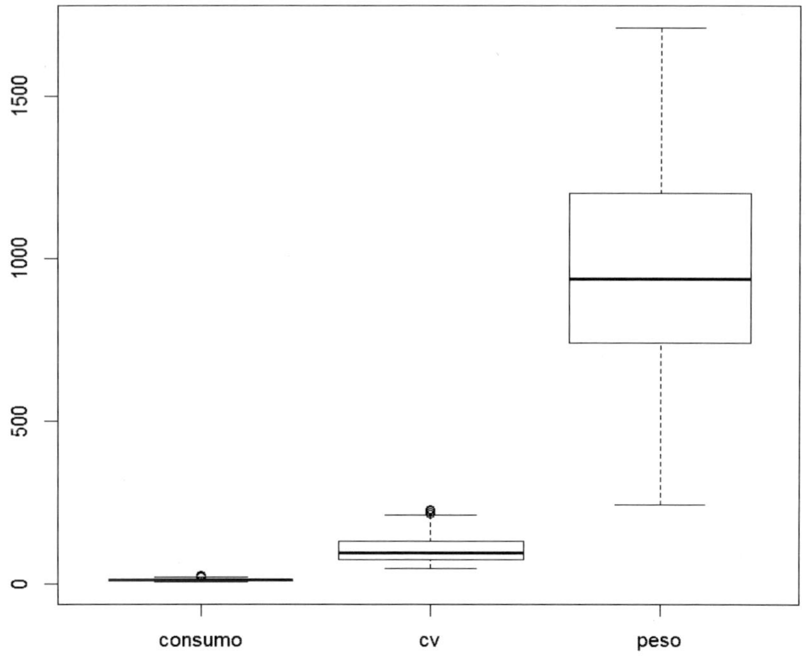

Figure 5-21

You can also make a multiple scatter plot (Figure 5-22) for the first two variables of the dataframe using the following syntax:

```
plot(data[,1:2])
```

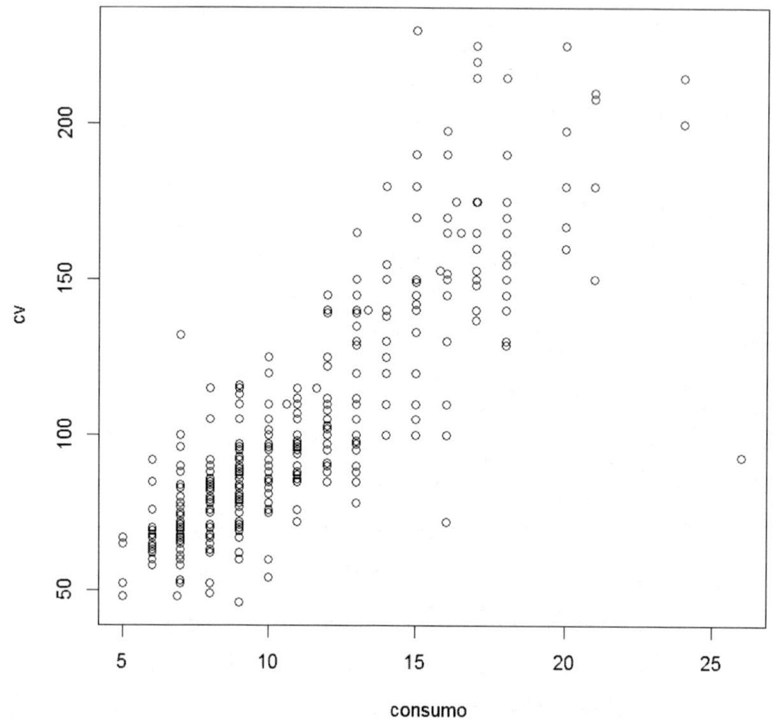

Figure 5-22

Next, we make a scatter plot of consumption versus power for all levels of the origin factor (Figure 5-23). The syntax would be as follows:

```
> coplot(consumption~cv|origin)
```

Next we make a box and whiskers graph of the consumption for all levels of the origin factor (Figure 5-24). The syntax would be as follows:

```
> boxplot(consumption~origin)
```

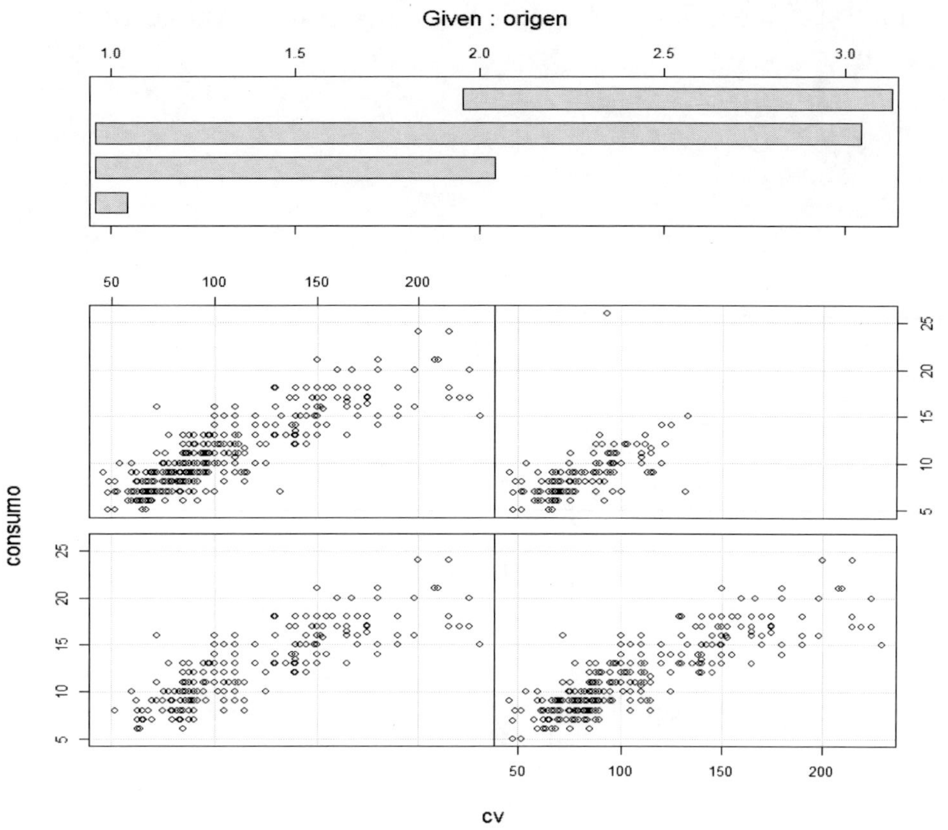

Figure 5-23

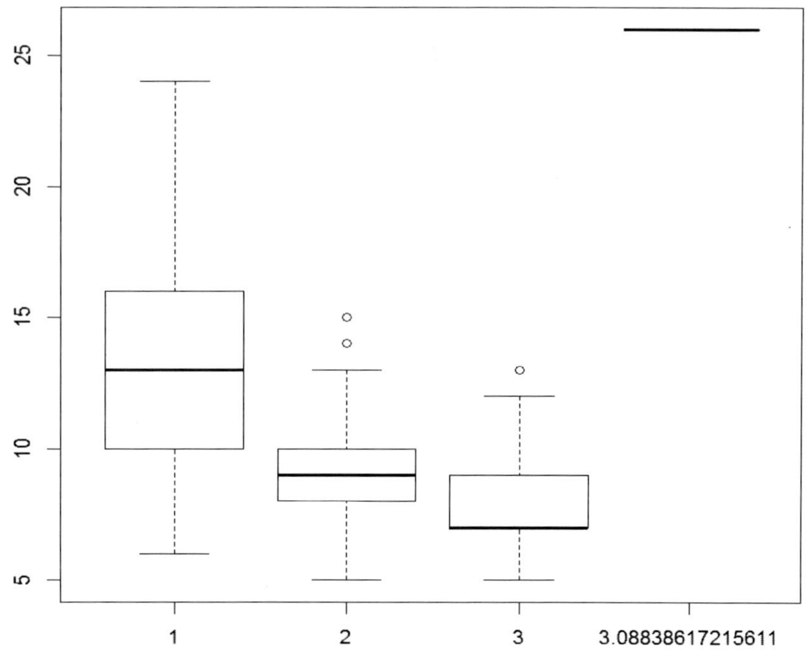

Figure 5-24

In Figure 5-25 is the bar graph for the *source* variable built through the following syntax:

```
> frequencies=table(origin)
> frequencies
source
 1    2    3
253 73 80
> barplot(frequencies)
```

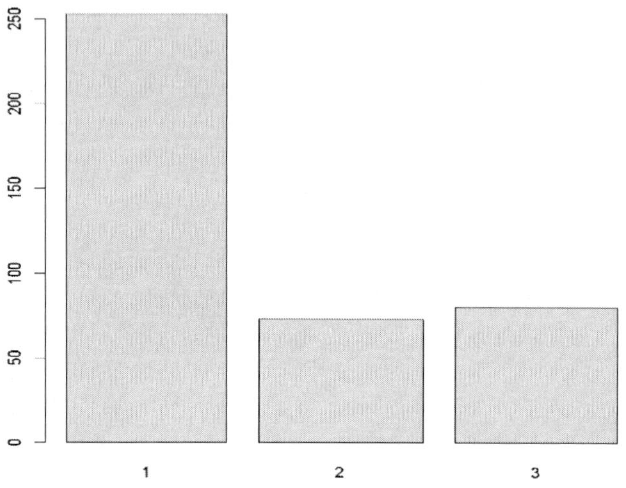

Figure 5-25

In Figure 5-26 is the graph of sectors for the *source* variable built through the following syntax:

```
> foot(frequencies, c( USA, Europe, Japan)
```

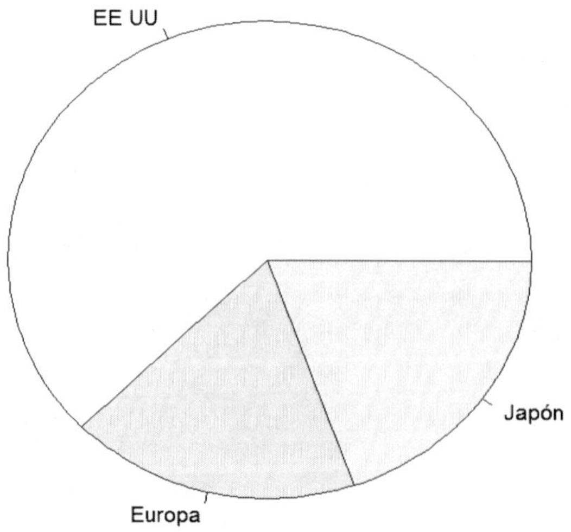

Figure 5-26

If we want to represent the consumption kernel density function (Figure 5-27) we will use the following syntax:

```
> density(consumption)
```

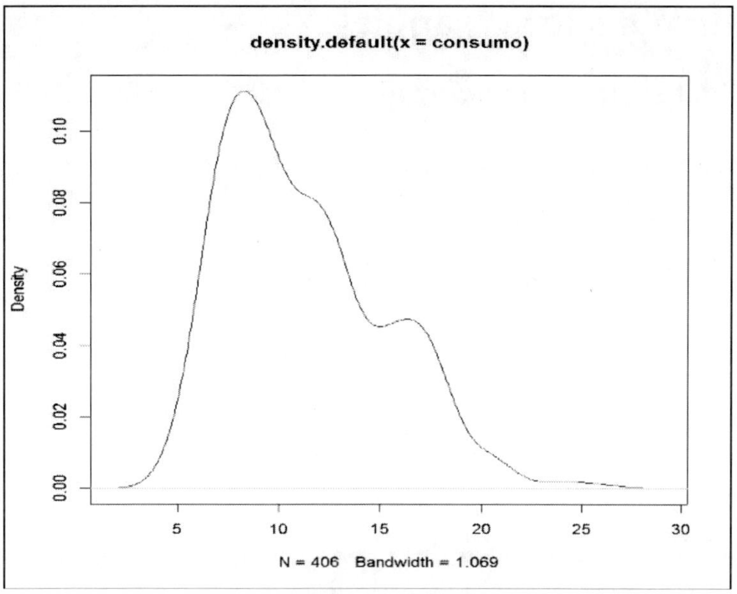

Figure 5-27

If we want to represent the multiple scatter diagram of the variables *consumption, cv* and *engine* (Figure 5-28) we will use the following syntax:

```
> pairs(~consumption+cv+engine)
```

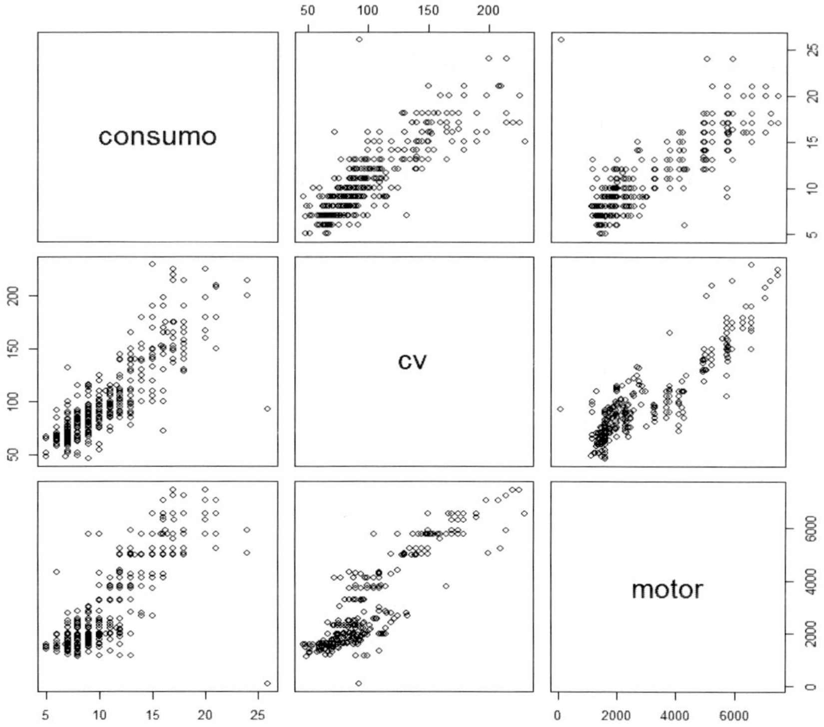

Figure 5-28

5.3 COMMANDS FOR EXPLORATORY GRAPHICS IN R. ADVANCED SYNTAX

5.3.1 Box and Whisker Graphics

The general syntax of the *boxplot* command is as follows:

```
boxplot(x, ..., range = 1.5, width = NULL, varwidth = FALSE, notch = FALSE,
    names, boxwex = 0.8, data = parent.frame(), plot = TRUE, border =
    par("fg"), col = NULL, log = "", pars = NULL, horizontal = FALSE, add =
    FALSE)
```

Their arguments are:

- *x, ...:* The variables that will be graficados with the boxplot. It can be a set of separate vectors, each corresponding to a boxplot component or a list containing such vectors. Alternatively, a especifiación of the form "x ~ g" can be given to indicate that the observations in vector "x" will be grouped according to the levels of the factor "g". In this case, the argument "data" can be used to allow for NA values in the data.

- *range*: Determines the extension of the whiskers on the box. If positive, the whiskers extend to the most extreme data, which is no more than the interquartile range of the box. A value of zero causes the whiskers to extend to the extreme data.

- *width*: A vector that gives the relative widths of the boxes.

- *varwidth*: If "TRUE", the boxes are drawn with widths proportional to the square roots of the number of observations in the groups.

- *notch:* If "TRUE", a wedge is drawn on each side of the boxes. When the cu˜nas of two gr'cu˜nas of box do not overlap, then the medians are significativamente different at a level of 5%.

- *names*: A group of labels to be printed under each boxplot.

- *boxwex*: A scale factor to be applied to all boxes. When there are only a few groups, the appearance of gr'gr´afico can be improved by making the boxes narrower.

- *data*: "data.frame", "list", or "environment" in which the names of the variables are evaluated when "x" is a formula.

- *plot*: If it is "TRUE" (default) then the gráfico is produced. Otherwise, summaries of the boxplots are produced.
- *border*: An optional vector of colors for the box lines and must be equal to the gr'number n´umero.
- *col*: If especifica, the colors it contains will be used in the body of the boxes.
- *log*: To indicate if the coordinates "x" or "y" or will be graficadas in logarithmic scale.
- *pars*: The parameters gráficos can be passed as arguments for the boxplot.
- *horizontal*: If "FALSE" (default) the boxplots are drawn vertically.
- *add*: If "TRUE" add a boxplot to the current gráfico.

The following syntax is also supported for the boxplot command:

boxplot.formula(formula, data = NULL, subset, na.action, ...)

The arguments are as follows:

- *formula*: A formula such as "y ~ x".
- data: A data.frame (or list) from which the variables in "formula" will be taken.
- *subset*: An optional vector for especificar a subset of observations to be used in the fitting process.
- *na.action*: A function which indicates what should happen when data contains NA.

As an example, we will represent the box and whiskers graph (Figure 5-29) of the consumption of cars according to displacement using boxes with green notches, with the boxes are drawn with widths proportional to the square roots of the number of observations in the groups and with titles for the graph and axes.

```
> library(foreign)
coches=read.spss("G:/R/data/coches.sav", use.value.labels=TRUE)
> attach(cars)
```

```
> boxplot(consumption~cylinder, notch=TRUE,varwidth=TRUE,col=
"green", main="Automotive data", xlab="Number of cylinders",
ylab="consumption")
```

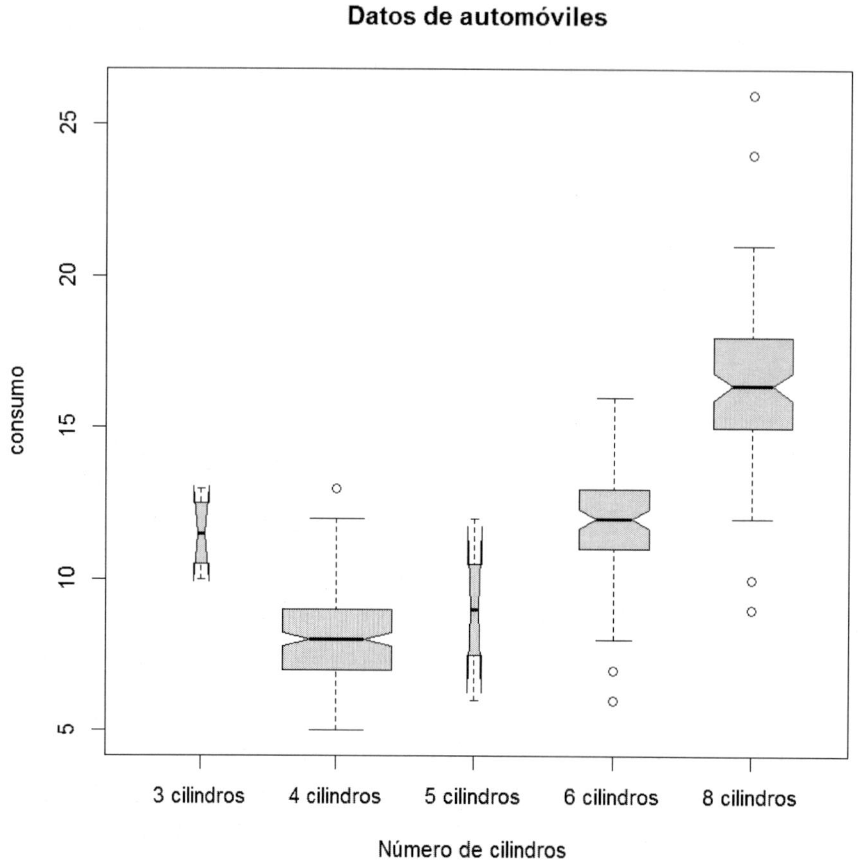

Figure 5-29

Below is the previous box and mustache graph in horizontal and without the special features that had been provided (Figure 5-30).

```
> boxplot(consumption~cylinder, horizontal=TRUE)
```

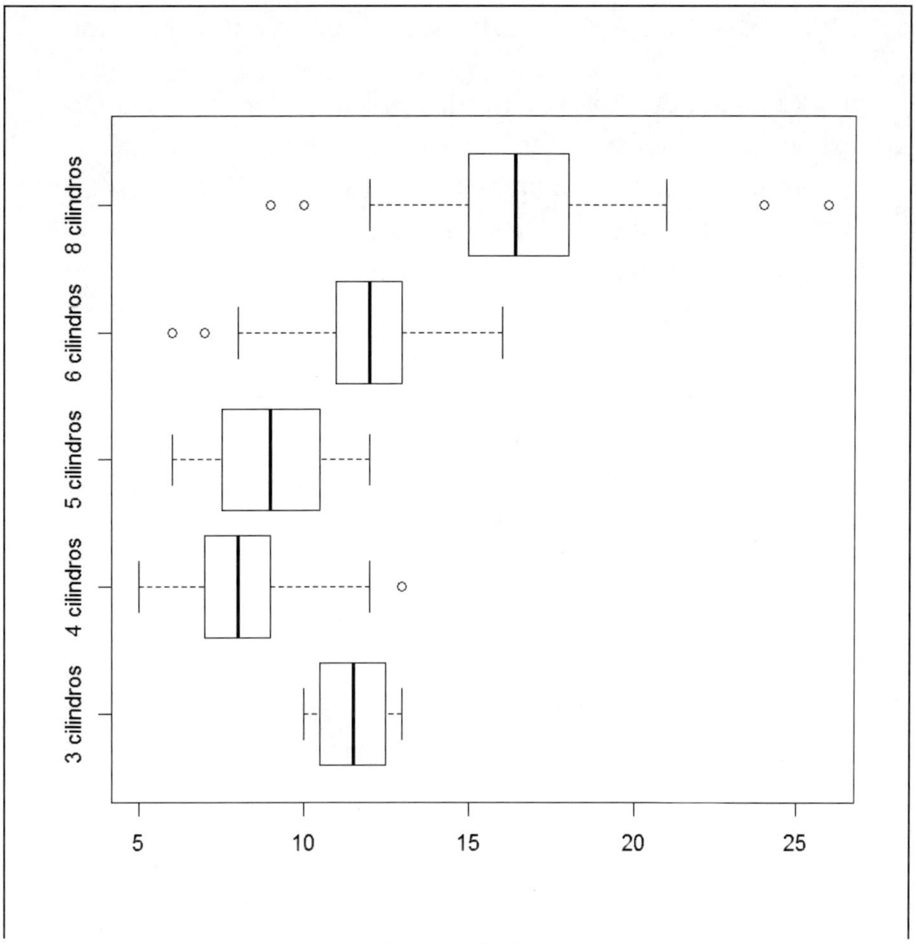

Figure 5-30

5.3.2 Bar graphs

The barplot command has the following general syntax:

barplot(height, width = 1, space = NULL, names.arg = NULL, legend.text = NULL, beside = FALSE, horiz = FALSE, col = heat.colors(NR), border = par("fg"), main = NULL, sub = NULL, xlab = NULL, ylab = NULL, xlim = NULL, ylim = NULL, axes = TRUE, axisnames = TRUE, inside = TRUE, plot = TRUE, ...)

The arguments are as follows:

- *height*: Vector or matrix of values that describe the bars. If it is a vector, then gráfico corresponds to a sequence of rectangular bars whose heights correspond to the values of the vector. If it is an array and "beside=FALSE" then each bar in the gráfico corresponds to a column so that the values in each column will be represented by sub-stacked bars, but if "beside=TRUE" then the values in each column will be juxtaposed instead of stacked.

- *width*: It is optional, for especificar by means of a vector the width of the bars.
- *space*: For fijar the space between bars, as a fraction of the average bar width. You can especificarse with a single number or one number per bar. In the case that the represented object is an array and "beside=TRUE", this argument can be given by two numbers, the first for the space between bars of the same group and the second for the space between groups. By default it is set to 'c(0,1)' for the case of an array object and beside=TRUE.
- *names.arg*: A vector of names to be placed under each bar
- or group of bars. If omitted, then the names are taken from the name attributes that are contained in the object especificado in "height".
- *legend.text*: A text vector to build a legend for gráfico. It is only useful if "height" is an array, in which case the legends will correspond to its filas.
- *beside*: A logical value. If "FALSE" the "height" columns will be represented by stacked bars and if "TRUE" then they will be represented by juxtaposed bars.
- *horizon*: A logical value. If "FALSE" the bars are drawn vertically.
- *col*: For especificar a color vector for the bars.
- *border*: To indicate the color to be used on the borders of the bars.
- *main,* sub: For headline and subtitle the gr'gr´afico.
- *xlab*: To name the x-axis.
- *ylab*: To give name to the y-axis.
- *xlim*: To delimit the range of values on the x-axis.
- *ylim*: To delimit the range of values on the y-axis.
- *axes*: Logical argument. If it is "TRUE", draw the corresponding axis.
- *axisnames*: Logical argument. If it is "TRUE", and if you have especificado "names.arg", then the other axis is drawn and labeled.

- *plot*: Logical argument. If it is "FALSE" it does not occur gráfico

Let's suppose that we have collected a sample of 20 data that we code with values from 1 to 4.

First we define the variable *x that* contains the data vector.

```
> x=c(1,2,3,1,2,1,2,4,1,3,2,4,1,2,3,1,2,4,2,1)
```

Below we define the frequency table corresponding to this data:

```
> data=table(x)
> data
x
1 2 3 4
7 7 3 3
```

Finally, we created the bar chart with legends (Figure 5-31):

```
> barplot(data, main="Relative frequency", xlab="Variable val
ues")
```

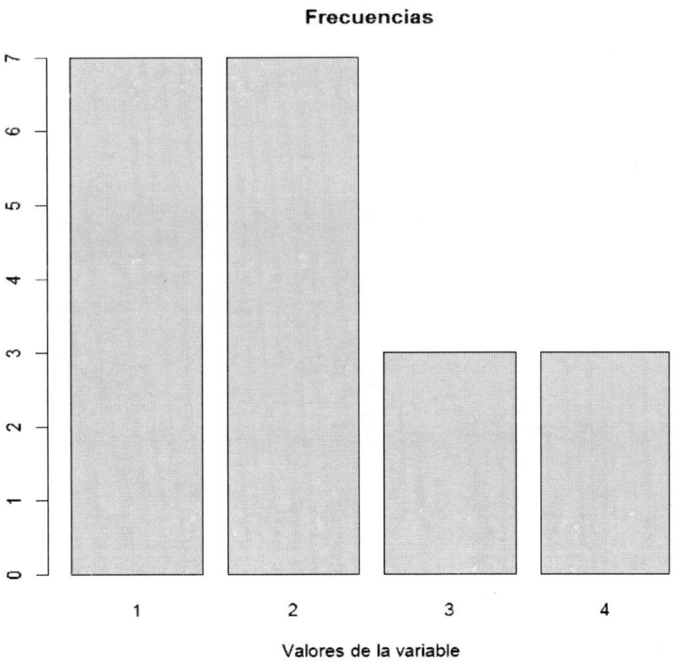

Figure 5-31

Then we assign names to the categories of the variable and perform the histogram horizontally (Figure 5-32).

```
> barplot(data, main="Relative frequency", xlab="variable val
ues", horizon=TRUE, names.arg=c("Under 18", "From 18 to 45",
"From 45 to 65","Over 65"))
```

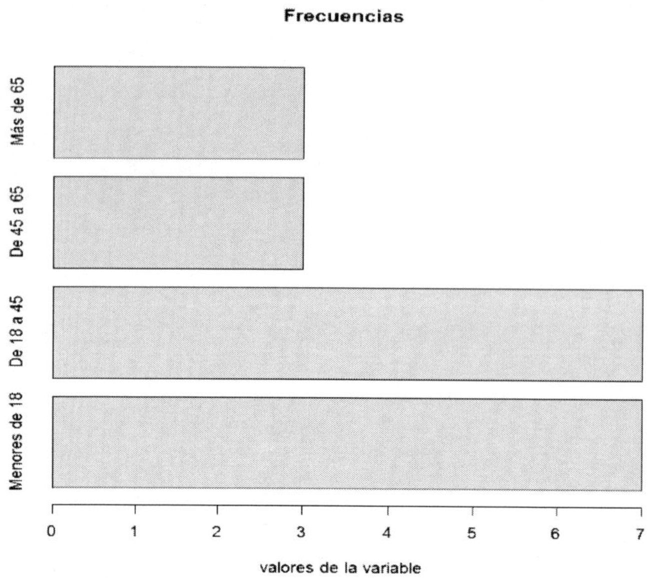

Figure 5-32

In the following example we represent a bar chart with the variables x,y and z grouped next to each other with the thickness 45,50 and 35 with the colors 1 and 2 (black and red) and with the legend man and woman (Figure 5-33).

```
> barplot(height = cbind(x = c(4, 9), y = c(8, 2), z = c(3, 7
) ), beside = TRUE, width = c(45, 50, 35), col = c(1, 2), leg
end.text = c("hombre", "mujer"))
```

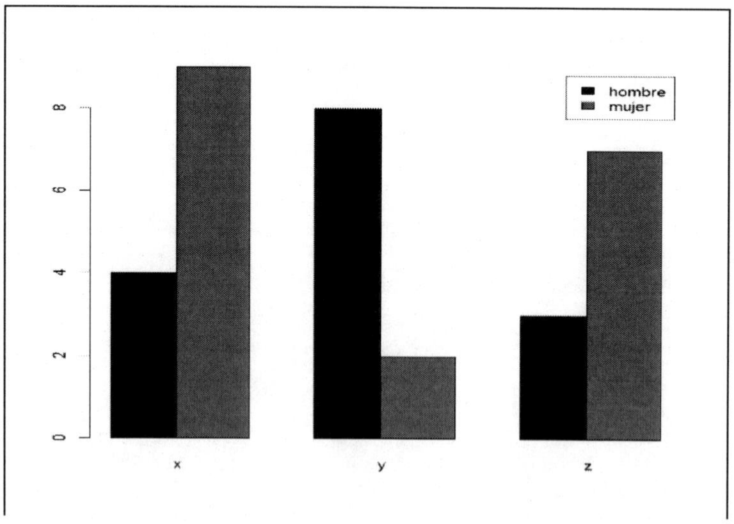

Figure 5-33

5.3.3 Sectorial graphics

The piechart command has the following general syntax:

piechart(x, labels=names(x), shadow=FALSE, edges=200, radius=0.8, col=NULL, main=NULL, ...)

Their arguments are as follows:

- *x*: Vector of positive quantities to be graphed as areas of the sectors.
- *labels*: A vector of character "strings" that give names to areas.
- *shadow*: A logical vector that indicates if a shadow effect will be applied for the gráfico, when using fill colors.
- *edges*: Approximates the circular outside linc by means of a polygon with the number of sides especificado, which by default is 200.
- *radius*: The sector is drawn centered in a square box whose sides move from -1 to 1. If long labels are used it may be necessary to use smaller radii.
- *col*: A vector of colors, to fill the sectors of the gráfico.
- *main*: To give title to gráfico.

We begin by making a simple pie chart for the data of the first example in the previous section (Figure 5-33).

```
> x=c(1,2,3,1,2,1,2,4,1,3,2,4,1,2,3,1,2,4,2,1)
> data=table(x)
> foot(data)
```

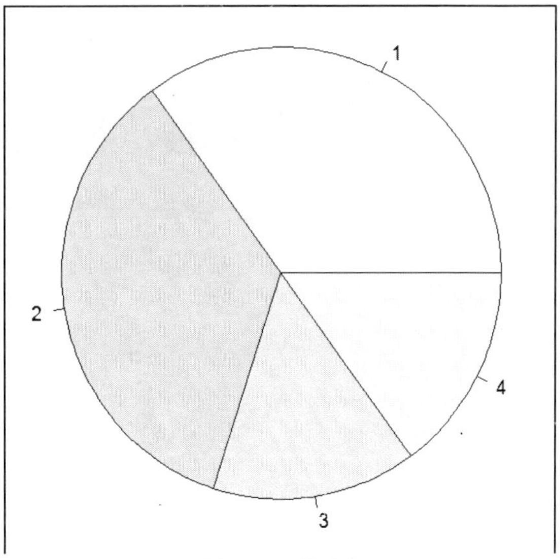

Figure 5-33

Now we place labels in the categories of the previous graph (Figure 5-33).

```
> tags=c("Under 18", "From 18 to 45", "From 45 to 65","Over 6
5")
> foot(data, labels=label)
```

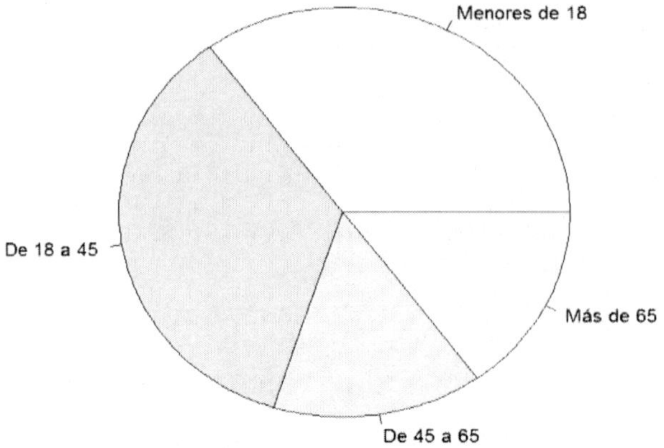

Figure 5-33

In the following example we consider a sample of synthetic indicators of population density in 5 countries of the world. The idea is to make a sector diagram for the densities with the label of each country and with indicative percentages of the population densities. In addition, each sector will be colored with rainbow colors.

```
> sample <- c(15, 12, 4, 16, 8)
> countries <- c( USA, England, Australia, Germany, France)
> pct = round(sample/sum(sample)*100)
> countries1=paste(countries,pct,"%")
> foot(sample,labels = countries1, col=rainbow(length(countri
es))
```

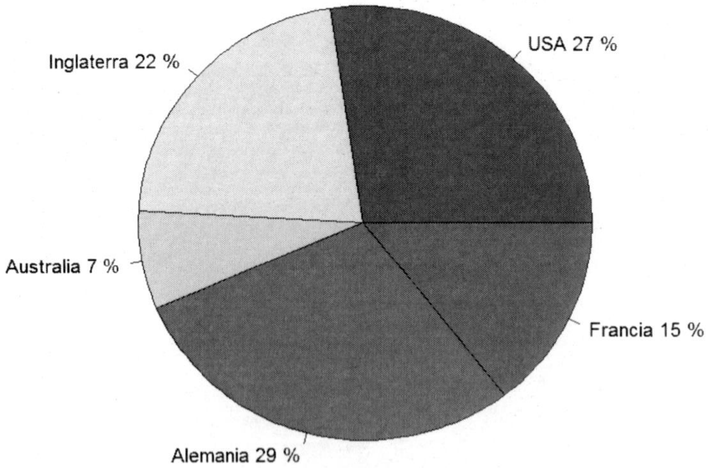

Figure 5-34

5.3.4 Histograms

The general syntax of the *hist* command is as follows:

hist(x, breaks, freq = NULL, probability = !freq,
include.lowest = TRUE,

```
right = TRUE, col = NULL, border = par("fg"),
main = paste("Histogram of" , xname),
xlim = range(breaks), ylim = NULL,
xlab = xname, ylab,
axes = TRUE, plot = TRUE, labels = FALSE,
nclass = NULL, ...)
```

- The arguments for this command are:
- *x*: Vector of values for which the histogram is constructed.
- *breaks*: It can be a value that indicates the approximate number of classes or a vector whose elements indicate the boundary points between classes or intervals.
- *freq*: logical argument; if especifica is set to "TRUE", the histogram will represent the absolute frequencies or data count in each class; if especifica is set to "FALSE", the relative frequencies will be represented. By default, this argument takes the value of "TRUE" as long as the intervals are of equal width.
- *probability*: Especifica an alias for "!freq", for S. compatibility
- *include.lowest*: Logical argument; if especifica is set to "TRUE", an "x[i]" equal to the "breaks" value will be *included in* the first bar, if the argument "right = TRUE", or in the last one if not.
- *right:* Logical argument; if it is "TRUE", the intervals are open on the left-closed on the right (a,b]. For the first class or interval if "include.lowest=TRUE" the smaller value of the data will be included in it. If it is "FALSE" the intervals will be of the form [a,b) and the argument "include.lowest=TRUE" will have the significado to include the "highest" one.
- *col*: For definir the color of the bars. By default, "NULL" produces bars without background.
- *border*: For definir the color of the borders of the bars.
- *plot*: Logical argument. The default is "TRUE", and the result is the gráfico of the histogram; if especifica is set to "FALSE" the result is a list of counts for each interval.
- *labels*: Logical argument or character. If especifica is set to "TRUE" place labels above each bar.

- *nclass*: Numerical argument (integer) For S compatibility, "nclass=n" is equivalent to "breaks=n".

- ...: Parameters gráficos in addition to "title" and "axis".

As a first example we can represent a histogram relative to 500 data from a normal distribution with 10 blue intervals with pink separation lines (Figure 5-35).

The syntax could be the following:

```
> x<-rnorm(500)
> hist(x, freq = FALSE, breaks = 9, col="lightblue", border="
yellow")
```

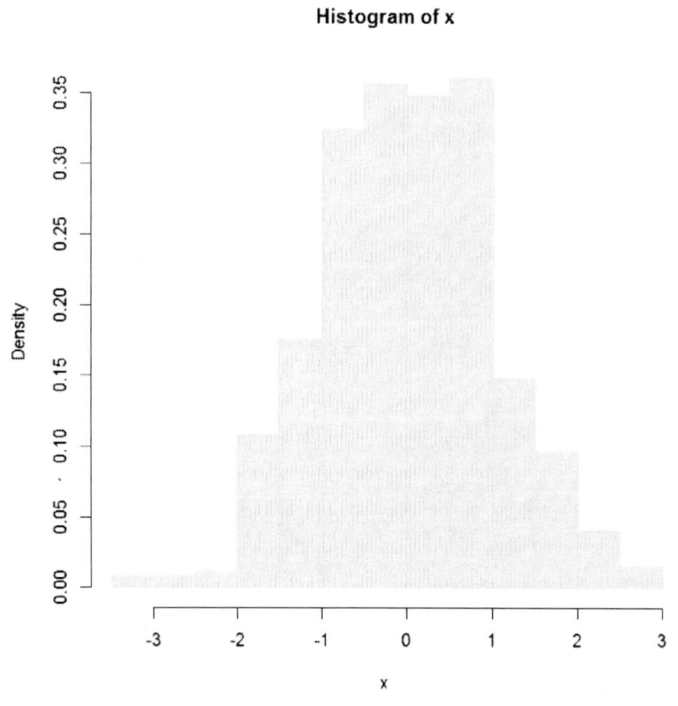

Figure 5-35

As a second example, let's consider the above histogram and add the setting to the Gauss bell (Figure 5-36).

The syntax would be as follows:

```
> x<-rnorm(500)
> hist(x,col="blue",freq=FALSE,xlim=c(-5,5))
> hist(x,col="lightblue",freq=FALSE,xlim=c(-5,5))
> curve(dnorm(x),-5,5,add=TRUE,col="blue")
```

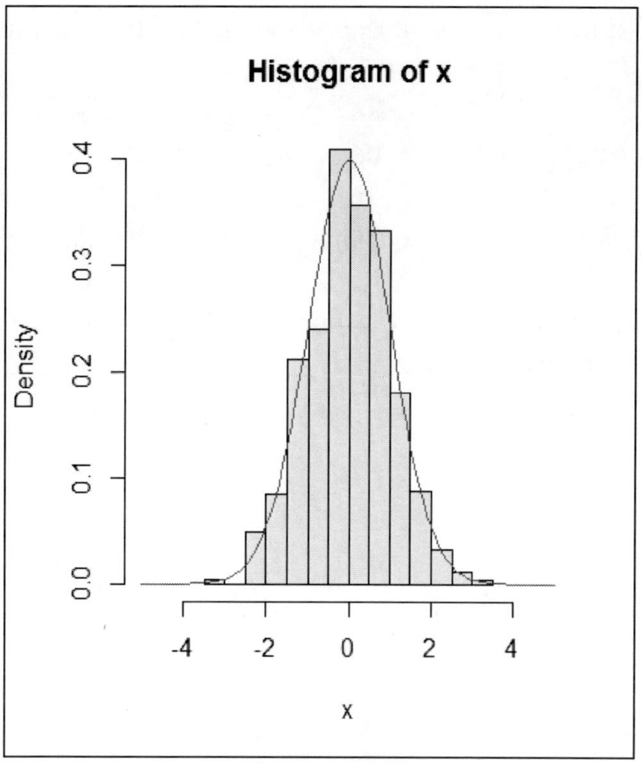

Figure 5-36

5.3.5 2.2.4 Gráficos of Points

The gráficos points are elegantly simple and allow numerous variations. In R you can get this gráfico by using the "stripchart" function:

```
stripchart(x, method="overplot", jitter=0.1, offset=1/3, vertical=FALSE,
        group.names, xlim=NULL, ylim=NULL, main="", ylab="", xlab="",
    pch=0, col=par("fg"), cex=par("cex"))
```

The arguments of this function are the following:

- *x*: Object that contains the data to be graficados. It can be a single vector or a list of vectors, each corresponding to a component of gráfico. You can also use especificar "x~g" to indicate that the observations in vector "x" will be grouped according to the levels of the factor "g".

- *method*: For especificar the method to be used to separate matching points. If especifica is set as "overplot" (default) the matching points are overlapped. If especifica is set to "jitter", they will be separated, and if especifica is set to "stack" the match points are stacked. You can specify the level of separation.

- *Vertical*: Logical argument. The default is "FALSE" and the points are grafican horizontally.

- *group.names*: Group of labels to be printed next to or under each gráfico

- ...: Additional gráficos parameters that can especificarse as arguments.

As a first example, we made a horizontal point graph for the values of the consumption variable according to car capacity with the coinciding points separated by a tenth (Figure 5-37)

```
> stripchart(consumption~cylinder, jitter=0.1, vertical=FALSE
)
```

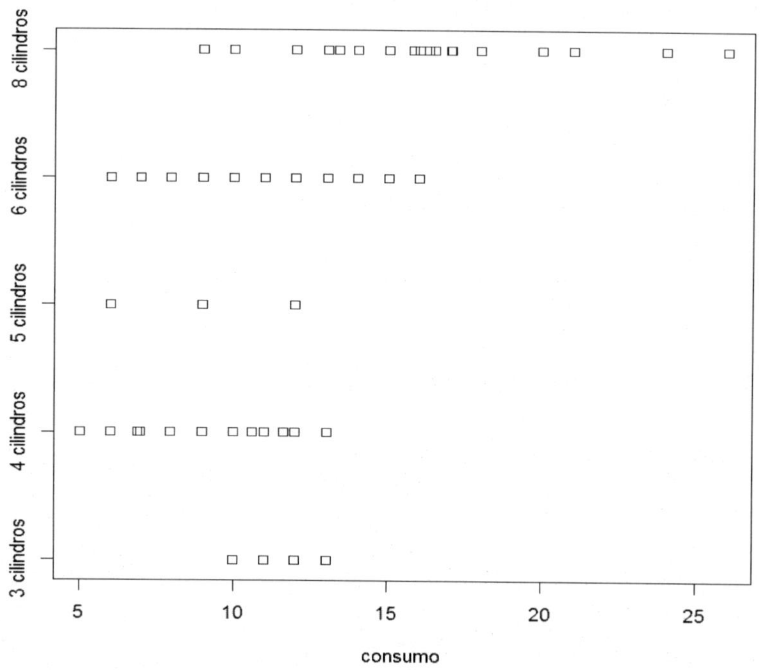

Figure 5-37

Another command that e usually uses in R for point graphics is dotchart, whose syntax is as follows:

dotchart(x, labels=, cex=, gcolor= , main=, xlab=

The dotted graph for the variable *x is* made, with the labels defined in *labels* with the label colors defined in *gcolor* and with the label size defined in *cex*. The *main* argument defines the title of the graph and the *xlab* argument defines the legend for the horizontal axis of the graph.

In the following example, a point graph is made for the variable of the file *zonemad.sav* that defines the occupied ones in the industry in the neighborhoods of Madrid (*ocuin),* using as label the neighborhood (variable *b)* and as size of label 0.7. Place an appropriate title on the horizontal axis (Figure 5-38).

First we read the SPSS format file and enable its variables.

```
> library(foreign)
> barrios=read.spss("G:/R/data/zonasmad.sav", use.value.label
s=TRUE)
> attach(neighborhoods)
```

Secondly, we look at the variables.

```
> summary(neighborhoods)
        Length Class Mode
b       18      -none- character
pt      18      -none- numeric
p14     18      -none- numeric
p65     18      -none- numeric
p10     18      -none- numeric
anal    18      -none- numeric
nes     18      -none- numeric
ocu     18      -none- numeric
ocuin 18      -none- numeric
ocuser 18      -none- numeric
tec     18      -none- numeric
dp      18      -none- numeric
tm      18      -none- numeric
```

Next we made the graph of points...

```
> dotchart(ocuin, labels=b, xlab="Occupied in the industry by
districts", cex=.7)
```

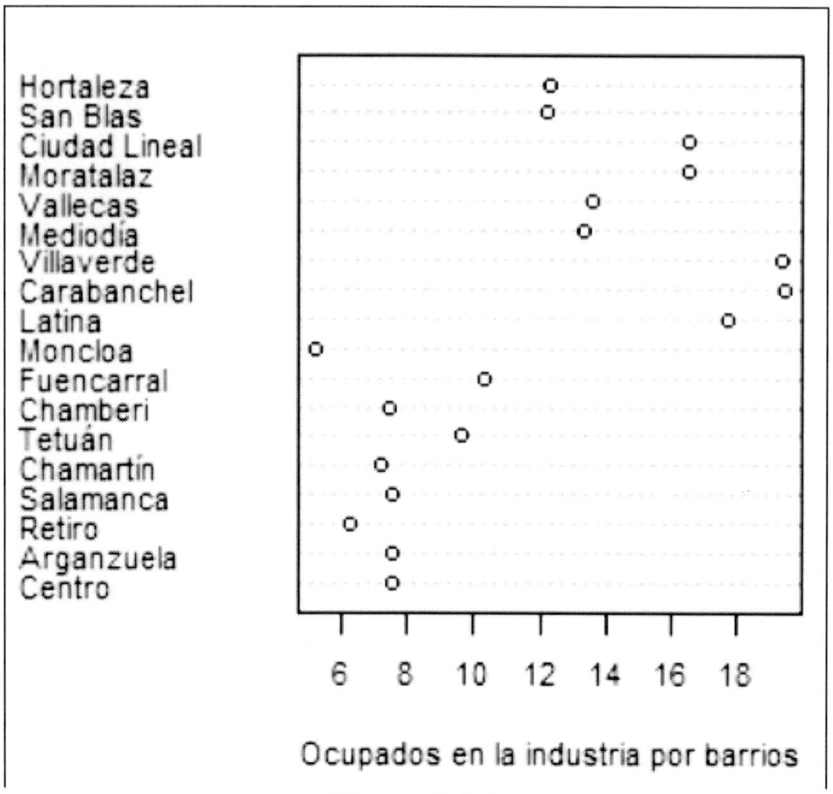

Figure 5-38

Exercise 1. Plotting a pie chart and bar graph for the car file source variable

The following syntax solves the problem and produces the graphs in Figures 5-39 and 5-40. The reading of the file is done with labels enabled for the qualitative variables in order to obtain the bar and sector labels correctly.

```
> library(foreign)
> coches=read.spss("G:/R/data/coches.sav", use.value.labels=T
RUE)
> attach(cars)
> data=table(origin)
> windows()
> foot(data)
> barplot(data)
```

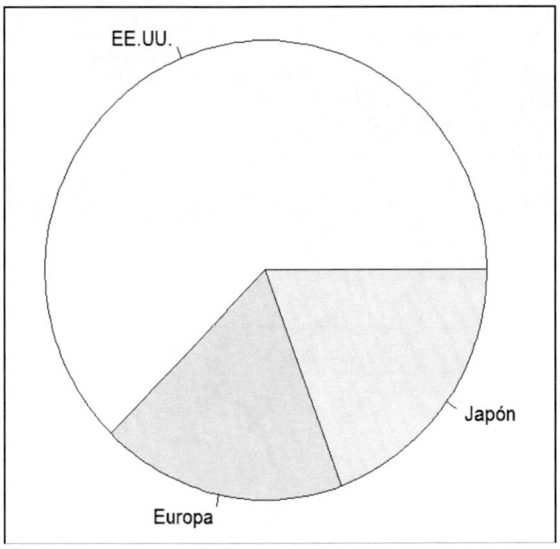

Figure 5-39

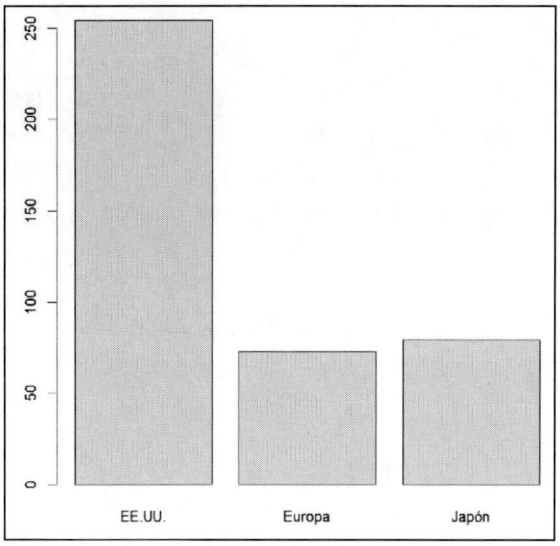

Figure 5-40

Exercise 2. Perform a graphic exploratory analysis of the variables of the viviendas.sav file that contains information about neighborhoods of a city, price of the houses, price of the land, value of the improvements and ratio between real price of kla vienda and appraised price.

First we read the file as a data frame and enable its variables for analysis.

```
> library(foreign)
> viviendas=read.spss("G:/R/data/viviendas.sav", use.value.la
bels=TRUE)
> attach(dwellings)
```

Next we explore the variables of the file and its nature.

```
> summary(housing)
          Length Class Mode
SC          2440    -none- character
neighborhood    2440    factor numeric
valterr 2440    -none- numeric
valmejor 2440    -none- numeric
valtot     2440    -none- numeric
price     2440    -none- numeric
rate        2440    -none- numeric
```

We observe that the variable SC is of type character, the variable neighborhood is a factor and the rest of the variables are numerical

We will make a QQ chart of normality for the qualitative variables and we will place it in a 2x2 graphic matrix (Figure 5-41).

```
> windows()
> pair(mfrow=c(2,2))
> qqnorm(valterr, xlab="land value")
> qqnorm(valmejor, xlab="value of improvements")
> qqnorm(price, xlab="house price")
> qqnorm(rate, xlab="ratio real price over assessed price")
```

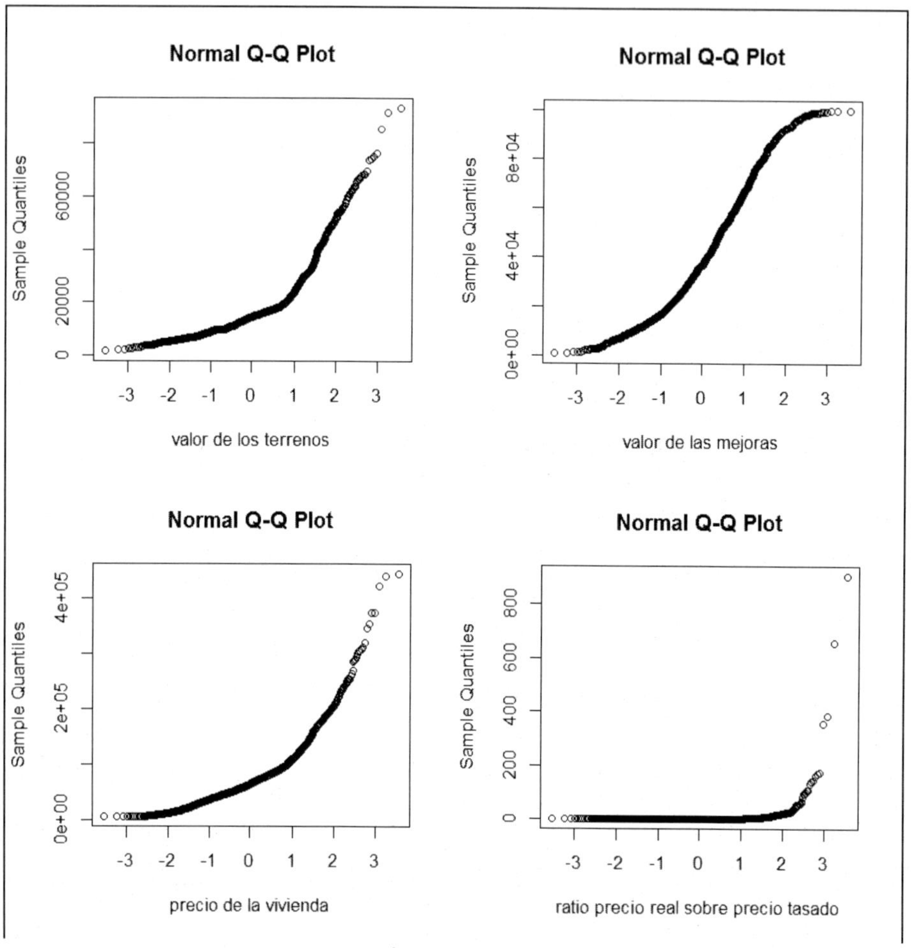

Figure 5-41

It is observed that the variables have difficulties in adjusting to normality since the point clouds do not adjust well to the diagonal of the first quadrant, except perhaps the value of the improvements.

Next we make the histograms of the four variables (Figure 5-42).

```
> windows()
> pair(mfrow=c(2,2))
> hist(valterr)
> hist(valmejor)
> hist(price)
> hist(rate)
```

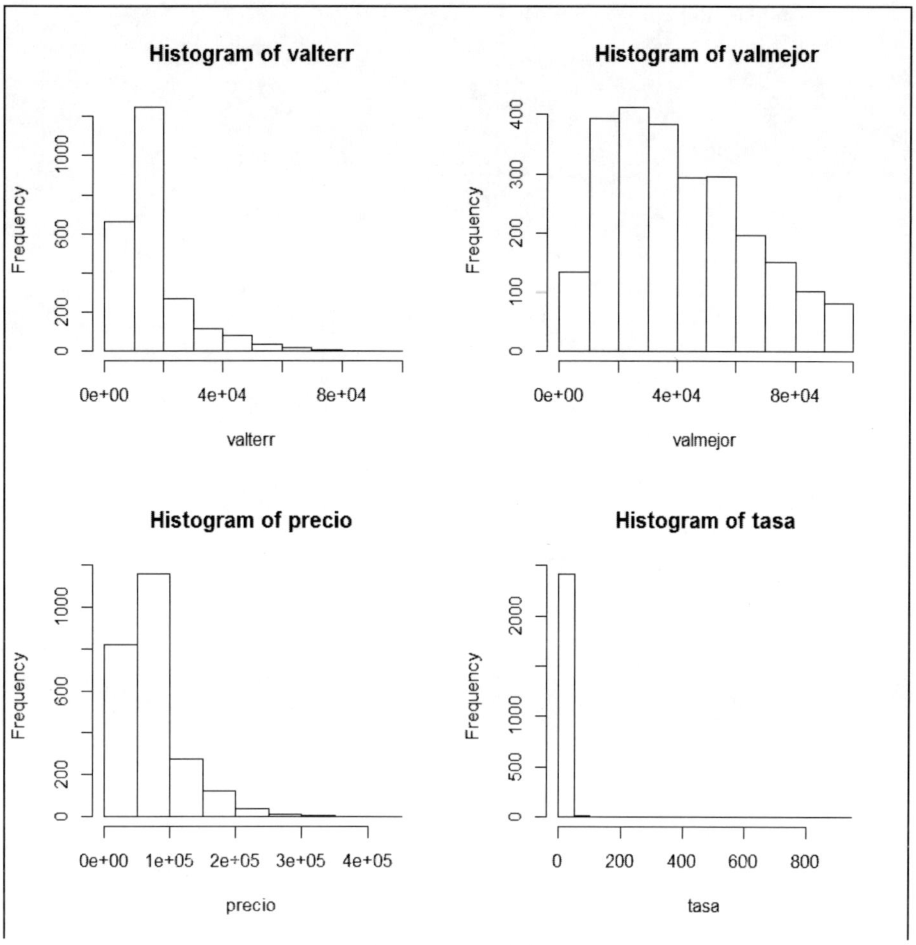

Figure 5-42

The histograms certify the lack of normality of the variables because they do not adjust to the bell of Gauss.

Below are the density graphs of the variables (Figure 5-43).

```
> windows()
> pair(mfrow=c(2,2))
> plot(density(valterr))
> plot(density(valmejor))
> plot(density(price))
> plot(density(rate))
```

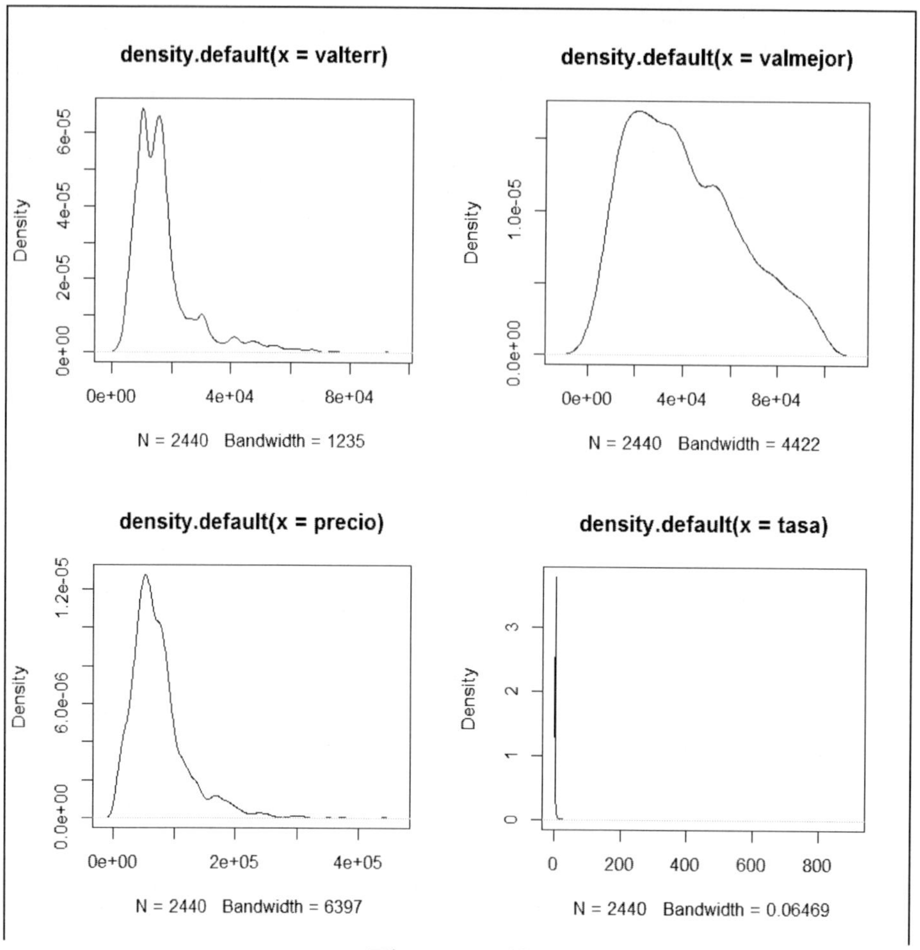

Figure 5-43

It is re-certified that the densities are not close to the ade Gauss bell, except for the value of the improvements.

Now we make the box and whiskers graphs of the four variables (Figure 5-44).

```
> windows()
> boxplot(data)
```

It is observed the lack of symmetry in the variables except for *valmejor*. The variables are asymmetrical on the left, which causes a lack of normality. The land value and housing price variables present higher atypical values, a result very much in line with the real estate bubble market.

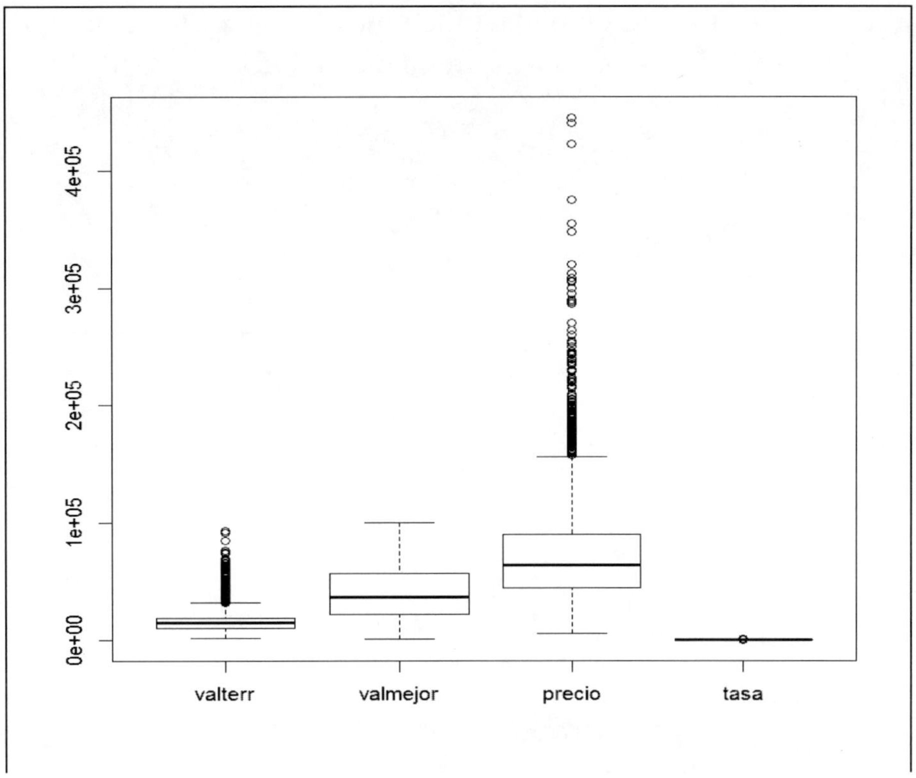

Figure 5-44

We can also make the multiple scatter plot (Figure 5-45).

```
> plot(data)
```

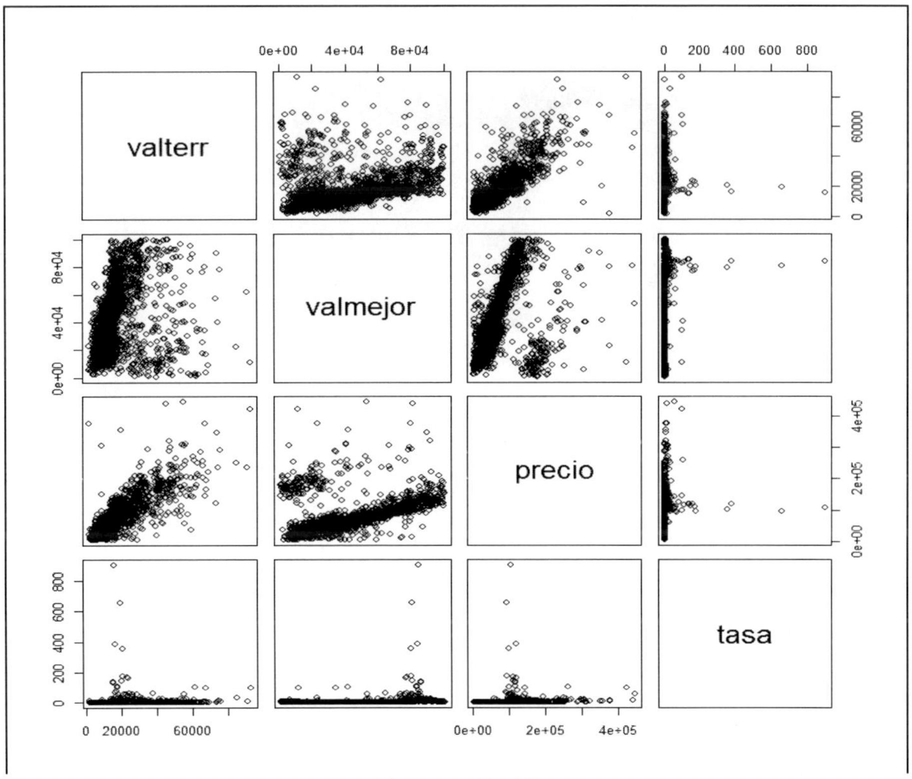

Figure 5-45

Now we make the bar graph and the sector graph of the qualitative variable *neighborhood.* Previously we built your frequency table

```
> table=table(neighborhood)
> table
neighborhood
 A    B    C    D    E    F    G
 42 319 258 467 500 372 482
> barplot(table)
> foot(table)
```

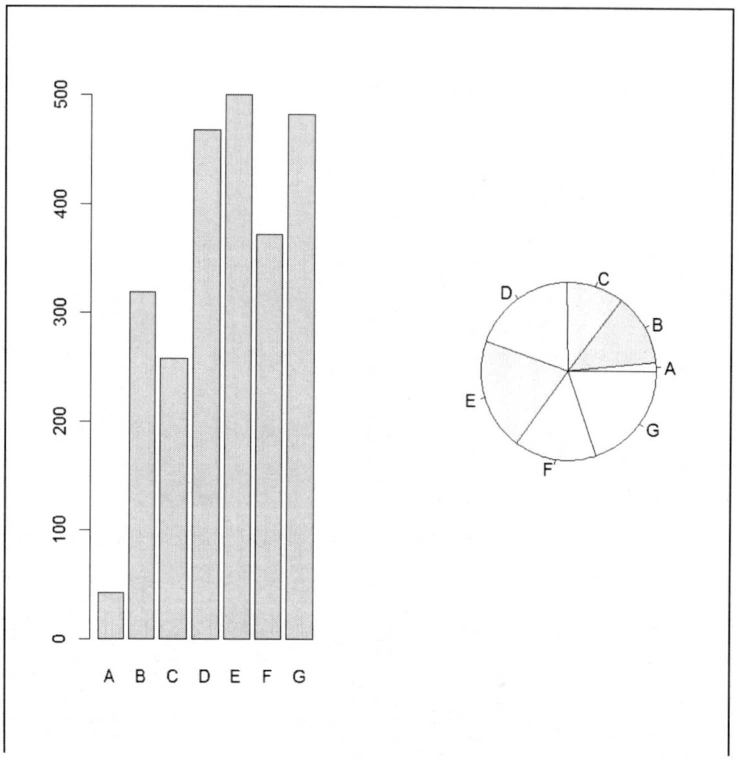

Chapter 6

DESCRIPTIVE STATISTICS THROUGH R

6.1 QUANTITATIVE EXPRESSION OF DISTRIBUTIONS

Descriptive statistics studies the different ways of summarizing the distributions of the variables by means of *position* (or centralization) *measurements, bearing* in mind the error made in the summary by means of the corresponding *dispersion measurements*. In turn, we will analyze the form of the distribution by means of *shape measures*. The frequency histogram already gave us a visual representation of the three most important properties of the sample data related to variables: the shape of their distribution, their central tendency and their dispersion. Now it is a matter of quantifying these concepts.

6.2 POSITION MEASUREMENTS: MEAN, MEDIAN, MODE, QUANTILES, PERCENTILES AND MOMENTS

The purpose of the position measurements is to find values that synthesize the frequency distributions. Instead of handling all the data about the characteristics or variables, a task that can be heavy, we can characterize their frequency distribution by means of some numerical values, choosing as a summary of the data a central value around which the values of the variable are distributed. The value of the variable chosen to represent a distribution is called average or position measurement, and is a representative value of all values that takes the variable. Must be between the highest and lowest value of the variable.

But these position measurements of a frequency distribution have to meet certain conditions in order to be truly representative of the variable they summarize. Any synthesis of a distribution will be considered as operative if each and every one of the values of the distribution are involved in its determination, being unique for each frequency distribution, and always being calculable and easy to obtain. The most common position measurements used in statistics are listed below.

Arithmetic mean: It is defined as the sum of all the values of the distribution divided by the total number of data. The mathematical expression that represents the arithmetic mean coincides with the first order moment with respect to the origin. But this is only valid in the simplest assumption, in which the data of the variable are ungrouped. In the case that we had a distribution with data grouped in intervals, the individual values of the variable would be unknown and, therefore, we could not use the previous formula. In this case, the data are grouped into classes, and it is hypothesized that the midpoint of the class interval (class mark) adequately represents the mean value of that class, so we would apply the original formula of the simple average for these values. In the case that the variable presents abnormally extreme values, these can distort the arithmetic mean, making it even unrepresentative. The statisticians who are not affected by the extreme values of the sample are called robust statisticians. The mean is not a robust statistic. As we will see later, this disadvantage is not possessed by the median. If the distribution of frequencies is $(xi,_{ni})$, where *xi is the* value of the variable or the class marks, and where *neither* the absolute frequencies nor the median are present, the arithmetic mean, which we will represent by $\overline{x}$, is defined as follows:

$$\overline{X} = \frac{1}{N}\sum_{i=1}^{k} x_i n_i$$

Weighted arithmetic mean: Particular case of the arithmetic mean, which appears when each value of the variable *xi is* given a weight or *wi,* different from the frequency or repetition $_{ni}$. In this case, not all the values of the distribution intervene with the same weight in the calculation of the mean. The amount of these weights defines the importance of each value of the distribution in the calculation of the mean. If the frequency distribution is $(xi,_{ni},wi)$, where *xi is the* values of the variable or the class marks, *nor the* absolute frequencies, and *wi is the weights,* the weighted arithmetic mean, which we will represent as *W,* is defined as follows:

$$W = \frac{\sum_{i=1}^{k} x_i n_i w_i}{\sum_{i=1}^{k} n_i w_i}$$

Geometric mean: Be a frequency distribution $(xi,_{ni})$. The geometric mean, which we will represent as *G, is* defined as the *N-th* root of the product of the *N* values of the distribution.

$$G = \sqrt[N]{x_1^{n1} x_2^{n2} \ldots x_k^{nk}}$$

The most frequent use of the geometric mean is to average variables such as percentages, rates, index numbers, etc.; that is, it is used in cases where the variable is assumed to have cumulative variations. In addition, when the variable takes at least a value $xi = 0$, then *G is* canceled, and if the variable takes negative values, there may be a range of particular cases in which it is also not determined due to the problem of the index roots pair of negative numbers.

Harmonic mean: Be a distribution of frequencies $(xi, _{ni})$ The harmonic mean, which we will represent as *H, is* defined as follows

$$H = \frac{N}{\sum_{i=1}^{k} \frac{1}{x_i} n_i}$$

Note that the inverse of the harmonic mean is the arithmetic mean of the inverses of the variable values. It is not advisable in variable distributions with small values. It is usually used to average variables such as productivities, speeds, times, yields, exchange rates, etc.

Quadratic mean: Be a distribution of frequencies $(xi, _{ni})$. The root mean square, which we will represent as *C, is* defined as follows

$$C = \sqrt{\frac{n_1 x_1^2 + n_2 x_2^2 + \ldots + n_k x_k^2}{N}}$$

The square mean is the square root of the arithmetic mean of the squares of the values of the variable.

Median: The median is defined as the value of the distribution, assuming that it is ordered from lowest to highest, which leaves to its left and right the same frequency of observations, that is, the value of the variable that occupies the central place, assuming an odd number of data. If the number of data were even, it can be said that there are two median values, and the arithmetic mean between them is taken as the median. You could also define the median as the value of the distribution whose accumulated absolute frequency is *N/2* ($N = \sum_{ni}$). For distributions grouped in intervals, and assuming that all values within the median interval $[L_{i-1}, L_i]$ are uniformly distributed along it, the median, *Me,* can be calculated by the expression:

$$M_e = L_{i-1} + \frac{\frac{N}{2} - N_{i-1}}{n_i} c_i$$

where N_{i-1}<N/2<Ni *and [*Li 1, L_i *is* the interval following the one containing N/2, being ci its amplitude.

The advantage of the median is that it is not influenced by extreme values (robust statistic). The median is very useful in process control charts.

Mode*:* Mode is the value of the variable that is repeated the most times, and consequently, in a frequency distribution, it is the value of the variable that is affected by the maximum frequency of the distribution. In distributions not grouped in intervals, the column of absolute frequencies is observed, and the value of the distribution to which the highest frequency corresponds will be the mode. Sometimes distributions of variables with more than one mode (bimodal, trimodal, etc.) appear, and even frequency distributions that have an absolute and a relative mode.

In distributions grouped in intervals of the same amplitude, we will not really have a modal value, but a modal interval (the interval with the highest frequency [L_{i-1}, L_i]). We can take as modal value the lower end of the interval (*M0* = L_{i-1}) or the upper end (*M0* = L_i), or make the mode equal to the class mark of the modal interval (*M0* = *xi*). If all the values of the modal interval are evenly distributed within it, the mode will be closer to that contiguous interval whose frequency is greater, with the distances from the mode, *M0,* to the contiguous intervals being inversely proportional to the frequencies of those intervals. The mode can be calculated using the expression:

$$M_0 = L_{i-1} + \frac{n_{i+1}}{n_{i-1} + n_{i+1}} c_i$$

For intervals of different amplitude, the frequency densities actually give us the number of values in each interval unit for each interval. The highest frequency density, now, determines the modal interval [L_{i-1}, L_i], calculating the mode through the expression:

$$M_0 = L_{i-1} + \frac{d_{i+1}}{d_{i-1} + d_{i+1}} c_i \text{ with } \quad d_i = n_i / c_i$$

It should be noted that mode is the most representative measure in the case of distributions of variables on a nominal scale.

This is because distributions of this type present data that cannot be sorted, so that for these distributions it is not possible to perform elementary operations with their observations. Mode is used especially when the values of the variable are highly concentrated towards a certain value. It will only be used in distributions with a high total frequency.

Quantiles: The *k-order* quantiles are defined as the values of the distribution, assuming that it is ordered from lowest to highest, which divide it into k-parts with the same frequency of observations. Therefore, there will be $r=k-1$ quantile of order k. The first quantile of order k leaves on the left the fraction 1/k of the frequency of observations. The second quantiles of order *k leaves on its left the fraction* 2/k of observation frequency. The *r-th quantile of* order *k leaves to its* left the *r/k fraction* (100r/k percent) of the observation frequency, and to its right it leaves the 1– *r/k fraction (*100*(k– r)/k percent) of the observation* frequency. The *r-enth* quantile of order *k is* usually designated as *Qr,k.*

For the more general case, in which the distribution is grouped in intervals, the *r-th* quantile of order *k will* take the value:

$$Q_{r,k} = L_{i-1} + \frac{\frac{rN}{k} - N_{i-1}}{n_i} c_i$$

where $N_{i-1} < rN/k < N_i$ *and [*Li 1, *Li-is* the interval following the one containing rN/k, being ci its amplitude.

For *k=4* and *r=1*,2,3, *we* will have the 3 quartiles. For *k=10* and *r=1*,2,...,9, we will have the 9 deciles, and for *k=100* and *r=1*,2,...,99, *we will* have the 99 percentiles. The *quartiles* are, therefore, the three values of the distribution that divide it in 4 equal parts, that is to say, in 4 intervals within each one it is included 25% of the values of the distribution. The *deciles are* the 9 points that divide the distribution into 10 parts, such that within each one 10% of the values of the distribution are included. *Percentiles are* the 99 points that divide the distribution into 100 parts, such that within each part 1% of the distribution values are included.

For distributions not grouped in intervals, the first quartile, *Q1*.4, is the value of the distribution that occupies the place *N/4; the* second quartile, *Q2*.4, is the value of the distribution that occupies the place 2N/4; the third quartile, *Q3*.4, is the value of the distribution that occupies the place 3N/4; the *r-th* decile, *Qr.*10, is the value of the distribution that occupies the place *rN/10* (*r=1*.2,....9); and the *rth* percentile, *Qr,*100, is the value of the distribution that occupies the place *rN/100* (*r=1*,2,...,99).

Potential moments : Among the values that characterize a distribution of frequencies, we have the potential moments. Two distributions that have their equal moments are equal, and they will be more similar the closer their moments are. The moments are usually considered with respect to the origin and with respect to the average.

The moment of order *r with* respect to the origin is defined as

$$a_r = \frac{1}{N}\sum_{i=1}^{k} x_i^r n_i$$

The order moment *r with* respect to the average is defined as

$$m_r = \frac{1}{N}\sum_{i=1}^{k} (x_j - \bar{x})^r n_i$$

There is a general formula that relates the moment of order *r centered in the* mean *mr and* the moments centered in the origin $a_r, a_{r-1}, \ldots, a_1$:

$$m_r = a_r - \binom{r}{1} a_{r-1} a_1 + \binom{r}{2} a_{r-2} a_1^2 - \cdots + (-1)^r a_1^r$$

It is observed that the moment of order 1 centered in the origin *a1* coincides with the arithmetic mean of the distribution, and the moment of order 2 centered in the mean m_2 coincides with the variance of the distribution.

Relationship between averages and Foster's formula : It is demonstrated that for the same frequency distribution (x_i, n_i) it is verified that *Harmonic mean* ≤ *Geometric mean* ≤ *Arithmetic mean,* that is to say

$$H \le G \le \bar{X}$$

To relate all the averages we have Foster's formula, which says that the average of *order m* has the following expression:

$$M_{(m)} = \sqrt[m]{\frac{n_1 x_1^m + n_2 x_2^m + \ldots + n_k x_k^m}{N}} = \left(\frac{1}{N}\sum x_i^m n_i\right)^{1/m}$$

so that as particular cases of Foster's formula we obtain the harmonic mean, the geometric mean, the arithmetic mean and the quadratic mean by doing $m = -1$, $m = 0$, $m = 1$ and $m = 2$ respectively. We can write then that:

$$M_{(-1)} = H,\ M_{(0)} = G,\ M_{(1)} = \bar{X},\ M_{(2)} = C$$

The relationship between averages can be extended by introducing the *C-square* average as follows:

$$H \le G \le \bar{X} \le C$$

DISPERSION MEASURES

The dispersion measures allow to calculate the representativeness of a position measure, for which it will be necessary to quantify the distance of the different values of the distribution from that measure. Such a distance is what, in statistical terms, we will call variability or dispersion of the distribution. The purpose of dispersion measures is to study to what extent, for a given frequency distribution, the central tendency or position measures are representative as a synthesis of all the distribution information. Measuring the representativeness of a position measure is equivalent to quantifying the separation of the distribution values from that measure. For example, if we want to study to what degree an arithmetic mean marks a generalizable central tendency of the behavior of all the elements of the studied set, we will have to pay attention to the separation or deviation of each value with respect to the mean. If all the values are close to the mean value, it will be representative of them.

The greater or lesser separation of the values of one distribution from another, which is intended to be its synthesis, is called dispersion or variability. It will be, then, so much more representative the arithmetic mean of a variable the more grouped around it are the averaged values; on the contrary, it will be so much more reprehensible, for not being representative, the greater dispersion of the variable's values with respect to the mean.

It is therefore necessary to complete the information that can be deduced from a measure of position or centralization, accompanied by one or more coefficients that measure the degree of dispersion of the distribution of the variable with respect to that measure of centralization. These coefficients are what we call measures of dispersion. Initially, we distinguish between absolute and relative measures of dispersion, understood as relative measures that do not depend on the units of measurement. Later, the absolute and relative measures are classified according to whether they are measures referred to averages or not.

Among the measures of absolute dispersion not referred to averages we have the *path,* or difference between the highest and lowest value of a distribution, and the *interquartile path,* or difference between the third quartile and the first. Among the relative dispersion measures not referred to averages we have the *opening coefficient, or quotient between the highest value and the lowest value of* a distribution, and the *relative path, or quotient between the path and the* mean, as well as the *semi-quartile path, or quotient between the* interquartile path and the sum of the first and third quartiles.

Among the measures of absolute dispersion referred to averages we have the *mean deviations*, the *variance* and the *standard deviation*. These measures of dispersion involve averages, and allow us to measure the error we make by using the average in question as a summary of the data. As a simpler measure of dispersion relative to the measure of position *P,* we could consider the deviations of each value to the average and average these deviations, ie, consider the value $D = (\sum xi - P)_{ni/N}$, $i=1...k$. But this, which would be the first thing that would occur to us, has as serious disadvantage the possible compensations of the positive deviations with the negative ones when carrying out the sum, being able to obtain a small measure being the big dispersion. To solve this inconvenience, the absolute values of the deviations are considered, or they are squared. Then, the most interesting measures of dispersion are defined, among which we have:

Mean deviations*:* To measure the effectiveness of the mean, the mean *deviation from the arithmetic mean is* considered. The arithmetic mean is defined as the arithmetic mean of the absolute values of the differences between the values of the variable and the arithmetic mean, and its expression is the following:

$$D_m = \frac{1}{N}\sum_{i=1}^{k} | x_j - \bar{x} | \ n_i$$

To measure the effectiveness of the median, Me, *is* usually considered the *mean deviation from the median,* which is defined as the arithmetic mean of the absolute values of the differences between the values of the variable and the median, and whose expression is the following

$$D_{Me} = \frac{1}{N}\sum_{i=1}^{k} | x_j - M_e | \ n_i$$

Variance, quasi-variance, standard deviation and standard error: Of all the absolute measures of dispersion with respect to the arithmetic mean, the variance and its square root (the standard deviation), are the most important. If instead of considering the absolute values of the deviations from the average we consider their squares, there is a new measure of dispersion called variance, which we define as the arithmetic mean of the squares of the deviations of the variable's values to the arithmetic mean, that is, the second-order moment with respect to the arithmetic mean. It is defined by the expression:

$$\sigma^2 = \frac{1}{N}\sum_{i=1}^{k} (x_j - \bar{x})^2 n_i$$

As most important properties of the variance we have that it can never be negative, that it is equal to the second order moment with respect to the origin minus the first order moment squared, that if in the distribution of frequencies we add to all the values of the variable a constant the variance does not vary (a change of origin in the variable does not affect the variance), and that when multiplying the values of a distribution of frequencies by a constant *k the variance* is multiplied by the square of the constant.

Just as the mean deviations are expressed in the same units of measure as the distribution, the variance is not, since it will be given in the co-responding units, but squared. This makes it difficult to interpret, and makes it necessary to define the standard deviation or standard deviation. The standard deviation is the square root, with a positive sign, of the variance, so its expression will be:

$$\sigma = \sqrt{\frac{1}{N}\sum_{i=1}^{k}(x_j - \bar{x})^2 n_i}$$

Since it is the square root of the variance, it will be expressed in the same units of measurement as the distribution, which makes it more suitable as a measure of dispersion.

A widely used statistic as a measure of dispersion, especially due to its sampling properties, is quasi-variance, whose expression is:

$$S^2 = \frac{1}{N-1}\sum_{i=1}^{k}(x_j - \bar{x})^2 n_i$$

It is also considered the quasi-standard deviation, whose expression is

$$S = \sqrt{\frac{1}{N-1}\sum_{i=1}^{k}(x_j - \bar{x})^2 n_i}$$

Another statistic widely used as a measure of dispersion, also because of its sampling properties, is the standard error, whose value is $e=S/\sqrt{n}$.

It is always true that $DMe < Dm <$.

Among the measures of absolute dispersion referring to averages, we could have also defined the mean deviation from the mode and the quadratic deviations from the median and from the mode, which are given in the same units of measurement as the distribution, and which mark the representativeness of the averages to which they relate.

Among the relative dispersion measures (non-dimensional values that are not affected by the measurement units and are always specified as a ratio) used to compare position measures or averages, we have the *dispersion index with respect to the median* and the *Pearson variation coefficient.*

Pearson's Coefficient of Variation: Used to solve the problem of comparing arithmetic means of various distributions that can come, in general, in different units.

It is defined as the ratio of the standard deviation to the arithmetic mean ($V =$ $/ \sigma\bar{x}$). Obviously, the lower the coefficient of variation (V), the better the mean. We observe that when we make the quotient we eliminate the units, and therefore V is dimensionless. On the other hand, V represents the number of times it contains a $\bar{x}$

σ, and it is clear that the greater the number of times V contains a , $\sigma\bar{x}$ then, relatively, the greater the value of V *the* lesser the representation of $\bar{x}$. This coefficient is also usually expressed in as many percent as $V = 100\,(\sigma/\bar{x}\;)$.

As both in the calculation of$\sigma$and in the calculation of all the distribution values have intervened $\bar{x}$, V presents the guarantee, against other coefficients, that it uses all the distribution information. The lower value of V is zero, and the only problem case is when $\bar{x} = 0$, which would make V tend to infinity.

Dispersion index with respect to the median: Used to solve the problem of comparing medians of various distributions, which can come, in general, in different units. It is defined as the ratio of the mean deviation from the median to the arithmetic median $VMe = DMe\,/\,_{Me}$. Obviously, the lower the dispersion index, the better the median.

6.3 SHAPE MEASUREMENTS

Once the statistical analysis of information synthesis has begun, for which we have studied the measures of position and dispersion of the distribution of a variable, we need to know more about its behavior. We cannot base our conclusions only on expressions that are given in terms of measures of position and dispersion. While we try to globalize the behavior of the collective under study, for which measures of position are our best instrument, we should not proceed to an interpretation that implies a behavior of all elements of the collective uniformly constant and equal to the measure of position in question with an error given by the corresponding measure of dispersion. This error or disparity becomes more evident when analyzing the graphical representation of the distribution. Well, the measures of form of a distribution are based on its graphical representation, without getting to make the same.

Shape measurements are classified into *asymmetry measurements* and *kurtosis* or *pointing measurements*.

The purpose of the asymmetry measures is to elaborate an indicator that allows to establish the degree of symmetry (or asymmetry) that a distribution presents, without the need to carry out its graphic representation. Suppose that we have graphically represented a frequency distribution. If we draw a perpendicular to the abscissa axis $\bar{x}$ and we take this perpendicular as the symmetry axis, we will say that a distribution is symmetric if there is the same number of values on both sides of that axis, equidistant from $\bar{x}$ two to two, and such that each pair of equidistant values has the $\bar{x}$ same frequency. Otherwise, the distributions will be asymmetric.

Kurtosis measurements study the distribution of frequencies in the central area of the same. The greater or lesser concentration of frequencies around the mean and in the central zone of the distribution will result in a more or less targeted distribution. For this reason, kurtosis measures are also called pointing or central concentration measures. Kurtosis measurements are applied to bell shaped distributions, i.e. unimodal symmetric or slightly asymmetric. To study the kurtosis of a distribution, it is necessary to previously define a standard distribution, which we will take as a reference model. This distribution is the normal one, which corresponds to very common phenomena in nature, and whose graphic representation is a bell of Gauss.

Taking the normal as a reference, we will say that a distribution can be more targeted than the normal (i.e. *leptocurtic),* or less targeted (i.e. *platystic*). The normal distribution, from the point of view of kurtosis, is called *mesocurtic*. With kurtosis, we study the deformation, in a vertical direction, with respect to the normal, of a distribution. Next, the most common measures of asymmetry are defined, among which the following stand out

Fisher's asymmetry coefficient: Now we try to find a measure that picks up the symmetry or asymmetry of a distribution. If the distribution is symmetric, the symmetry axis of its graphic representation will be a line parallel to the ordinate axis, which passes through the point whose abscissa is the arithmetic mean. Therefore, when the distribution is asymmetric, we will refer the distribution values to this mean. If a distribution is symmetric, there is the same number of values to the right of $\bar{x}$, and therefore the same number of deviations with a positive sign as with a negative sign, being the sum of positive deviations equal to the sum of negative ones. We can start, then, from the deviations ($xi-\bar{x}$) elevated to an odd power in order not to lose the signs of the deviations. The simplest thing would be to take as a measure of asymmetry the average of these deviations, elevated to the simplest odd power (which is three); that is, we would take as a measure of asymmetry the moment of order three centered on the average. But, if we did this, this measure would be expressed in the same units as those of the variable but raised to the cube, so it is not invariant to a change of scale.

To get a dimensionless indicator, we must divide the previous expression by a quantity that comes in the same units of measure. This quantity is the cube of the standard deviation. In this way we obtain the asymmetry coefficient of R. A. Fisher, whose expression is

$$g_1 = \frac{m_3}{\sigma^3} = \frac{\frac{1}{N}\sum_{i=1}^{k}(x_j - \bar{x})^3 n_i}{\left(\frac{1}{N}\sum_{i=1}^{k}(x_j - \bar{x})^2 n_i\right)^{3/2}}$$

If *g1=0*, the distribution is symmetric; if *g1>0, the distribution is* positive asymmetric (right); and if g1<0, *the* distribution is negative asymmetric (left). The distribution is asymmetric to the right or positive when the sum of the positive deviations of its values from the mean is greater than the sum of the deviations with a negative sign (the graph of the distribution has more density to the right of the mean). Otherwise, the distribution is asymmetrical to the left or negative.

Standardized Fisher asymmetry coefficient: For *N>150, the* asymmetry coefficient is asymptotically normal of mean zero and variance 6/N. This fact leads us to consider the standardized asymmetry coefficient, whose expression is

$$g_s = \frac{g_1}{\sqrt{\frac{6}{N}}}$$

This coefficient is asymptotically normal (0.1).

Pearson's asymmetry coefficient: Karl Pearson proposed for campaniform, unimodal and moderately asymmetric distributions, the coefficient defined as $Ap = (\bar{x} - Mo)/\sigma$, where *Mo* is the mode. As $\bar{x}$ in a symmetric bell distribution = *Mo* = *Me,* if the distribution is positive or right-handed asymmetric, $\bar{x}$ it moves to the right of the mode, $\bar{x}$ –and therefore, $\bar{x} - Mo > 0$. In the case of negative asymmetric distribution, the mean is below Mo, *so the* value Mo $\bar{x}$ – < *0.* So we will have that if *Ap* = *0* the distribution is symmetric, if *Ap>0 the distribution is* positive asymmetric, and if Ap<0 the *distribution is* negative asymmetric. Also Pearson checked empirically for this type of distributions that 3($\bar{x}$ –*Me*) *Mo*$\approx\bar{x}$ –fulfilled (the median is always between the mean and the mode in moderately asymmetric distributions). For this reason, some authors use as Pearson's asymmetry coefficient the value *Ap* 3≈($\bar{x}$ – *Me*) $/\sigma$.

Bowley's asymmetry coefficient: It is based on the position of the quartiles and the median, and is given by the expression *Ab* = (*C3* + *C1* 2Me–) / (*C3* + *C1*). It is fulfilled that if *Ab=0* the distribution is symmetric, if *Ab>0 the distribution is* positive asymmetric, and if Ab<0 the *distribution is* negative asymmetric. *C1* and *C3* are the first and third quartiles respectively.

Absolute Asymmetry Coefficient: It is also based on the position of the quartiles and the median, and is given by the expression:

A = [(*C3 C2*–) –*C2 C1*–] / *S* = (*C3* + *C1* 2C2–) / *S* = *C3* + *C1* 2Me–/*S*

If *A* = 0 the distribution is symmetric, if *A>0 the distribution is* positive asymmetric, and if A<0 the distribution is negative asymmetric. *C1*, *C2* and *C3* are the quartiles of the distribution.

Once the asymmetry measures are presented, the most common kurtosis measures are defined below, among which the following stand out:

Kurtosis Coefficient: In the normal distribution, it is verified that $m4 = 3\sigma^4$, being *m4* the moment of order 4 with respect to the mean, and $\sigma$the standard deviation. If we consider the expression $g2 = m4/\sigma^4$ –, its value will be zero for the normal distribution. Therefore, the expression is used as a pointing or kurtosis coefficient:

$$g_2 = \frac{m_4}{\sigma^4} - 3 = \frac{\frac{1}{N}\sum_{i=1}^{k}(x_j - \bar{x})^4 n_i}{\left(\frac{1}{N}\sum_{i=1}^{k}(x_j - \bar{x})^2 n_i\right)^2} - 3$$

A distribution is *mesocurtic* (pointing equal to normal) when $g2 = 0$, is *leptocurtic (pointing* greater than normal) if $g2 > 0$, *and* is platystic *(pointing* less than normal) if $g2 < 0$.

Standardized kurtosis coefficient: For $N > 150$, *the* kurtosis coefficient is asymptotically normal from mean zero and variance 24/N. This fact leads us to consider the standardized kurtosis coefficient, whose expression is

$$g_{ks} = \frac{g_2}{\sqrt{\frac{6}{N}}}$$

This coefficient is asymptotically normal (0.1).

6.4 LINEAR CORRELATION COEFFICIENT BETWEEN TWO VARIABLES. SIMPLE LEAST SQUARED REGRESSION

The degree of mutual dependence between two variables is called correlation. The correlation *coefficient* tries to measure the intensity with which two variables are related. This concept is directly related to the concept of the regression curve. Through simple least squares regression, the functional structure of the relationship between two variables is expressed by adjusting the point cloud given by the pairs of values of the two variables to a curve in the best possible way (minimizing the variance of the error). The adjustment will be in the form $Y=f(x)+e$ or $X=f(Y)+e$, where e denotes the error committed whose variance should be minimal. The correlation coefficient measures the quality of that adjustment.

When the curve is a straight line, the regression is called linear, and in this case the correlation coefficient is called linear *correlation coefficient, and* measures the degree of linear association that exists between the variables. The adjustment will be in the form $Y=a+bX+e$ (regression line of *Y over X*), or $X=c+dY+e$ (regression line of *X over Y),* where $a = \bar{y} - \bar{x}\, S_{xy} / S_x^2$, $b = S_{xy} / S_x^2$, $c = \bar{x} - \bar{y}\, S_{xy} / S_y^2$, and $d = S_{xy} / S_y^2$.

Parameters *a* and *b are* called *regression coefficients of Y over X,* and parameters *c and d* are called *regression coefficients of X over Y*. You can also express the regression lines of *Y* over *X* and *X* over *Y* respectively of the form and $y - \bar{y} = (x - \bar{x}) S_{xy} / S_x^2 \;\; (x - \bar{x}) = (y - \bar{y}) S_{xy} / S_y^2$.

If we assume the adjustment of the form *Y=a+bX+e* (*regression line of Y over X*), the least squares criterion considers that the function that best fits the data is the one that minimizes the variance of the error *e,* which is equivalent to minimizing:

$$S(a,b) = \sum_{i,j} e_{ij}^2 = \sum_{i,j} (y_j - (a + bx_i))^2$$

Deriving from the parameters *a* and *b and* equaling zero we have:

$$\left.\begin{aligned} \frac{\partial S(a,b)}{\partial a} &= 2\sum_{i,j} (y_j - (a + bx_i))(-1) = 0 \\ \frac{\partial S(a,b)}{\partial b} &= 2\sum_{i,j} (y_j - (a + bx_i))(-x_i) = 0 \end{aligned}\right\} \Rightarrow \left\{\begin{aligned} \sum_j y_j &= na + b\sum_i x_i \\ \sum_{i,j} y_j x_i &= a\sum_i x_i + b\sum_i x_i^2 \end{aligned}\right.$$

The solutions *a* and *b of* this *system of normal equations* are $a = \bar{y} - \bar{x}\, S_{xy} / S_x^2$ and $b = S_{xy} / S_x^2$, which makes the regression line of Y over X:

$$Y = a + bX = \bar{y} - \bar{x}\, S_{xy} / S_x^2 + S_{xy} / S_x^2 \, x \Rightarrow y - \bar{y} = (x - \bar{x}) S_{xy} / S_x^2$$

Reasoning in a similar way, the regression line of *X* over *Y is* obtained.

The expression of the linear correlation coefficient between the variables *X* and *Y is* given by the expression:

$$r = \frac{\sigma_{xy}}{\sigma_x \sigma_y} = \frac{\sum_{i=1}^{h} \sum_{j=1}^{k} (x_i - \bar{x})\,(y_j - \bar{y})\, n_{ij}}{\sqrt{\sum_{i=1}^{h} (x_i - \bar{x})^2 \sum_{i=1}^{h} (y_i - \bar{y})^2}}$$

If *r = 1* there is a perfect positive correlation, and the functional relationship between both variables is exact and positive, varying both in the same direction (increasing one increases the other, and decreasing one decreases the other). If *r* =

–perfect negative correlation, and the functional relationship between both variables is exact and negative, varying both in the opposite direction (as one increases, the other decreases, and as one decreases, the other increases). If $r = 0$ the correlation is null, and the variables are not associated, being impossible to find a functional relation between them.

If $0<r<1$ *the* correlation is positive, but the degree of association between the two variables will be greater as r gets closer to 1, and will be less as r gets closer to zero. If $-1<r<0$ *the correlation is* negative, but the degree of association between the two variables will be greater as r gets closer *to* 1, and will –be less as r gets *closer to* zero.

The square of the correlation coefficient *r2,* generally denoted by *R2,* is called the *determination coefficient and* represents the percentage of variability of the dependent variable that is explained by the regression. Given its definition, *R2* can be expressed in a general way as a function of the variance of *Y and*(σ_Y^2) *the* residual variance as σ_e^2 follows

$$R^2 = \frac{Varianza\ de\ la\ regresión}{Varianza\ de\ Y} = \frac{\sigma_Y^2 - \sigma_e^2}{\sigma_Y^2} = 1 - \frac{\sigma_e^2}{\sigma_Y^2} = 1 - \frac{\sum_{i=1}^{h}\sum_{j=1}^{k}(y_i - \overbrace{(a - bx_i)}^{\hat{y}_i})^2 n_{ij}}{\sigma_Y^2}$$

In general, this is a measure of the goodness of the regression adjustment. If *R2 is* close to unity the adjustment is good and if *R2 is* close to zero the adjustment is bad. This definition and interpretation of *R2 is* valid for any type of adjustment even if it is not linear.

As for the relationship between correlation and independence, it is observed that by defining the linear correlation coefficient as $r = Sxy /(Sx\ Sy)$, if the variables are independent they will be unrelated, since $r=0$ because *Sxy* is zero when there is independence. Now, the reciprocal is not necessarily true, since two variables can be linearly uncorrelated and dependent, since being $r=0$, the only thing we can say is that the linear association is zero, but those variables can depend according to another type of association (parabolic, exponential, etc.)

6.5 CORRELATION COEFFICIENT BY RANGES

It is very typical to consider, especially in qualitative variables, the correlation coefficient between the ranges of the variables' values. It is understood that the range of a value of a variable is the place that occupies this value in the total set of values of the variable, assuming an ordering from less to more. *Ai* and *Bi* are the different modalities of two qualitative variables *X* and *Y.* Let *xi* and *yi* be the ranges or order numbers that correspond to Ai and $_{Bi,\ supposedly}$ ordered these modalities, with the scale that is determined, and from less to more. Spearman's rank correlation coefficient for the qualitative variables X and Y is defined as the linear correlation coefficient of the variables whose values are xi and yi.

This value is used to measure the degree of association of the qualitative variables X and Y based on the agreement or disagreement of the rankings of their modalities. The range correlation coefficient is also used for quantitative variables, with the clarification that the degree of association obtained is not that of the values of the variables, but of the range classifications of those values. This coefficient is given by:

$$\rho = 1 - \frac{6\sum_i d_i^2}{N^3 - N}$$

where $_{di} = xi\ yi-$. This coefficient is also called ordinal correlation coefficient, and because it is a correlation coefficient it varies between –1 and 1. When the agreement between the ranges is perfect, then $_{di}\,\rho - xi\ yi - \rho 0$ and $=1\rho$.

6.6 COVARIANCE

Sometimes it is necessary to introduce measures that refer to the joint distribution of two variables.

The *two-dimensional moment of order* (r,s) *centered in the origin,* is defined as:

$$a_{rs} = \frac{1}{N}\sum_{i=1}^{h}\sum_{j=1}^{k} x_i^r y_j^s n_{ij}$$

The *two-dimensional moment of order* (r,s) *centered on the means,* is defined as

$$m_{rs} = \frac{1}{N}\sum_{i=1}^{h}\sum_{j=1}^{k} (x_i - \bar{x})^r (y_j - \bar{y})^s n_{ij}$$

The moments of the marginal distribution of X *will* be calculated as $_{ar0}$ (those centered in the origin) and *mr0* (those centered in the mean). Specifically *a10* is the mean of X, and *m20* is the variance of X. In the same way, the moments of the marginal distribution of Y *will be* calculated as *a0s (those centered in the origin)* and *m0s* (those centered in the mean). Specifically, *a01 is the* mean of *Y,* and *m02 is the* variance of *Y.*

Among the measures related to the joint distribution, the most interesting one in any statistical analysis is the σ_{xy} covariance, which is defined as *m11* and represents the conjunctive variation of two variables. Its formula will be:

$$\sigma_{xy} = \frac{1}{N}\sum_{i=1}^{h}\sum_{j=1}^{k}(x_i - \bar{x})\ (y_j - \bar{y})\ n_{ij} = \frac{1}{N}\sum_{i=1}^{h}\sum_{j=1}^{k} x_i\, y_j n_{ij} - \overline{X}\,\overline{Y} = a_{11} - a_{10}\, a_{01}$$

Covariance is a measure of great importance because of its sign, rather than the value it can reach; the sign indicates the direction of the joint variation of the variables we are considering. Thus, if the covariance is positive, it means that both variables vary in the same direction around their means, while if the variation of the same takes place in the opposite direction, the covariance will take negative values.

Sometimes, quasi-variance is used because of its sampling properties. The expression of quasi-covariance is:

$$\hat{S}_{xy} = \frac{1}{N-1}\sum_{i=1}^{h}\sum_{j=1}^{k}(x_i - \bar{x})\ (y_j - \bar{y})\ n_{ij}$$

6.7 DISTRIBUTIONS OF MORE THAN TWO DIMENSIONS: COVARIANCE MATRIX AND CORRELATION MATRIX

When you have a three-dimensional *(X,Y,Z)* or enedimensional variable in general, you can also perform the description and analysis of the underlying frequency distributions. The complexity grows when the number of variables or factors that are analyzed simultaneously increases, but known the procedure for the three-dimensional case, its generalization to the *n-dimensional* is immediate. The analysis is equally valid when the characteristics analyzed are of a quantitative or qualitative nature.

As in two-dimensional distributions, the most usual form of representation of three-dimensional distributions are correlation tables (referring to quantitative characters) or contingency tables (referring to quantitative characters). As in the case of two-dimensional distribution, it is intended that the frequencies, both joint and marginal as well as conditioned, are easily located, and also their respective distributions.

An essential element in the study of enedimensional variables is the *covariance matrix*, which summarizes the covariances for all possible pairs of variables between *X1, X2, ..., Xn*. It is defined as:

$$C = \begin{pmatrix} \sigma_{11} & \sigma_{12} & \sigma_{13} & \cdots & \sigma_{1n} \\ \sigma_{21} & \sigma_{22} & \sigma_{23} & \cdots & \sigma_{2n} \\ . & . & . & \cdots & . \\ . & . & . & \cdots & . \\ \sigma_{n1} & \sigma_{n2} & \sigma_{n3} & \cdots & \sigma_{nn} \end{pmatrix}$$

where each σ_{ij} represents the covariance between *Xi* and *Xj* for all *i,j*.

The sign of each σ_{ij} indicates the direction of the joint variation of the two variables *Xi* and *Xj* that we are considering. If the covariance is positive, it means that both variables vary in the same direction, while if the variation of the same takes place in the opposite direction, the covariance will take negative values.

With the covariance matrix we simultaneously analyze the direction of the joint variation of all possible pairs of variables *Xi* and *Xj* for all *i,j*.

Another essential element in the study of enedimensional variables is the correlation *matrix*, which summarizes the correlations for all possible pairs of variables between *X1*, *X2*,...,*Xn*. It is defined as:

$$R = \begin{pmatrix} r_{11} & r_{12} & r_{13} & \cdots & r_{1n} \\ r_{21} & r_{22} & r_{23} & \cdots & r_{2n} \\ . & . & . & \cdots & . \\ . & . & . & \cdots & . \\ r_{n1} & r_{n2} & r_{n3} & \cdots & r_{nn} \end{pmatrix}$$

where each *rij* is the correlation coefficient between *Xi* and *Xj* for all *i,j*.

If given a series of variables *X1, X2, ..., Xn,* it is a matter of studying the degree of simultaneous dependence between all of them (or between groups of them), the correlation matrix can be used. If, on the basis of the intensity with which they depend, a function can be established that explains a variable by means of all the others, which are supposed to be its influential causes, we are faced with a problem of multiple regression, which will be studied in later chapters.

By means of the *multiple linear correlation coefficient the* degree of simultaneous linear association between all variables is studied, while by means of the simple *rij* correlation coefficients the degree of association between variables *Xi* and *Xj* is measured without taking into account the other variables.

6.8 BASIC R FUNCTIONS FOR DESCRIPTIVE STATISTICS

The basic module of R has a group of functions related to the descriptive statistics. The following table presents this type of functions.

sum(x)	Sum of the elements of the variable x
prod(x)	Product of the elements of the variable x
max(x)	Maximum value of the elements of the variable x
min(x)	Minimum value of the elements of the variable x
pmax(x,y,z..)	Vector of maximum values of variables x, y , z..
pmix(x,y,z..)	Vector of minimum values of the variables x, y , z..
cumprod(x)	Accumulated product vector of the elements of the variable x
cumsum(x)	Vector of accumulated sums of the elements of the variable x
cummax(x)	Vector of accumulated maximums of the elements of the variable x
cummin(x)	Vector of accumulated minima of the elements of vector A
range(x)	Minimum and maximum values of the elements of the variable x. range(x) = c(min(x), max(x))
rank(x)	Vector with the ranges of the elements of the variable x (place occupied by each element in x)
mean(x)	Arithmetic mean of the variable x
weighted. mean(x,y)	Weighted arithmetic mean of the variable x with weighting vector y
median(x)	median of variable x
quantile(x,probs=0.k)	k-th quantile of the variable x
IQR(x)	Interquartile range of the variable x
fivenum(x)	Minimum, first quartile, median, third quartile and maximum of the variable x
sd(x)	quasi-standard deviation of variable x
var(x)	Quasivariance of the variable x
cov(x)	Quasivariance of the variable x
var(x,y)	Covariance between variables x and y
cov(x,y)	Covariance between variables x and y
cor(x,y)	Correlation coefficient between variables x and y

var(M)	Covariance matrix for the columns of the variable matrix or data frame M
cov(M)	Covariance matrix for the columns of the variable matrix or data frame M
cor(M)	Correlation matrix for the columns of the variable matrix or data frame M
cor(M, use=, method=)	Correlation matrix for data frame M columns using the *pearson, spearman* or *kendall* methods in the *method* argument. The *use* argument is used for the treatment of missing values. Its value *all.obs* produces an error if there is missing, its value *everything* calculates correlations in the presence of missing values, its value *complete.obs removes* observations by list and its value pairwise.*complete* removes observations by pair.
diff(x)	Generates the first difference variable of the variable x
diff(x,lag=k)	Generates the variable k-that is the difference of the variable x
scale(*x*, center=TRUE, scale=TRUE)	Standardizes the vector or matrix x
summary(D)	Descriptive statistics of data frame D variables
apply(D, function)	Applies the function to all variables in the data frame D
table(x)	Frequency table of the variable x
table(x,y,z..)	Contingency table of the specified factors
xtabs(~x+y+z+...)	Contingency table of the specified factors

As a first example, we are going to read the file *coches.sav,* enable its variables and build a data frame with the quantitative variables to apply the descriptive statistics functions that encompasses the entire data frame.

```
> library(foreign)
> cars=read.spss("G:/R/data/coches.sav",use.value.labels=FALSE)
> attach(cars)
> variables=data.frame(consumption, hp, weight, engine, accel
)
> summary(variables)
      consumption              cv                weight               engine              acce
leration
 Min.       5.00    min.      46.00    Min.     244.0      Min.     66     Min.     : 8.00
 1st Qu.:   8.00    1st Qu.: 75.21     1st Qu.: 741.2      1st Qu.:1708   1st Qu.:13.62
 Medium : 10.31     Medium : 94.50     Medium : 936.5      Medium :2434   Medium :15.50
 Mean      :11.27   Mean     :104.45   Mean      : 989.5   Mean     :3180 Mean     :15.50
 3rd Qu.: 14.00     3rd Qu.:128.00     3rd Qu.:1203.8      3rd Qu.:4806   3rd Qu.:17.07
 Max.      :26.00   Max.     :230.00   Max.      :1713.0   Max.     :7456 Max.     :24.80
```

Next we calculate the average of all the variables of the data frame

```
> sapply(variables,mean)
   consumption          cv       weight       engine      acceleration
  11.27362            104.44688 989.51232   3179.73399      15.49507
```

Then we apply the function *fivenum* to all the variables of the data frame, that is, we calculate minimum (1), first quartile (2), median (3), third quartile (4) and maximum (5).

```
> sapply(variables,fivenum)
        cons     cv    weight engine acceleration
[1,]  5.0000 46.0    244.0      66.0       8.0
[2,]  8.0000 75.0    741.0    1704.0      13.6
[3,] 10.3061 94.5    936.5    2433.5      15.5
[4,] 14.0000 129.0   1204.0   4949.0      17.1
[5,] 26.0000 230.0   1713.0   7456.0      24.8
```

Then we calculate the correlation matrix and the covariance matrix of the data frame variables.

```
> cor(variables)
            consumption         cv        weight        engine
acceleration
consumption 1.0000000 0.8399707 0.8404614 0.8396855 -0.5004536
cv          0.8399707 1.0000000 0.8597706 0.8973697 -0.7022883
weight      0.8404614 0.8597706 1.0000000 0.9333371 -0.4151784
engine      0.8396855 0.8973697 0.9333371 1.0000000 -0.5446238
acel       -0.5004536-0.7022883-0.4151784-0.5446238  1.0000000

> cov(variables)
           consumption          cv        weight         engine  acceleration
consumption   15.587618   127.38870     939.9813     5715.405     -5.573838
cv           127.388701  1475.54956    9355.5868    59427.683    -76.101453
weight       939.981275  9355.58682   80245.9492   455817.077   -331.777470
engine      5715.404955 59427.68267 455817.0774  2972220.092  -2648.730449
acel          -5.573838   -76.10145    -331.7775    -2648.730      7.957951
```

We could also have used a matrix structure to calculate the matrices of previous variances and covariances.

We start by defining the matrix whose columns are the variables and then apply the functions to calculate the correlation and covariance matrices.

```
> variables1=cbind(consumption,cv,weight,engine,accel)
> cor(variables1)
            consumption         cv        weight        engine
acceleration
consumption 1.0000000 0.8399707 0.8404614 0.8396855 -0.5004536
```

```
cv            0.8399707 1.0000000 0.8597706 0.8973697 -0.7022883
weight        0.8404614 0.8597706 1.0000000 0.9333371 -0.4151784
engine        0.8396855 0.8973697 0.9333371 1.0000000 -0.5446238
acel         -0.5004536 -0.7022883-0.4151784 -0.5446238 1.0000000
```

```
> cov(variables1)
              consumption         cv       weight        engine          accel
consumption   15.587618    127.38870     939.9813      5715.405      -5.573838
cv           127.388701   1475.54956    9355.5868     59427.683     -76.101453
weight       939.981275   9355.58682   80245.9492    455817.077    -331.777470
engine      5715.404955  59427.68267  455817.0774   2972220.092   -2648.730449
acel          -5.573838    -76.10145    -331.7775     -2648.730       7.957951
```

Next, we calculate the frequency table of the *cylinder* variable and use it to construct a bar graph and a pie chart (Figures 6-1 and 6-2).

```
> table1=table(cylinder)
> table1
cylinder
 3    4    5    6    8
 4  207    3   84  108
> foot(table1)
> barplot(table1)
```

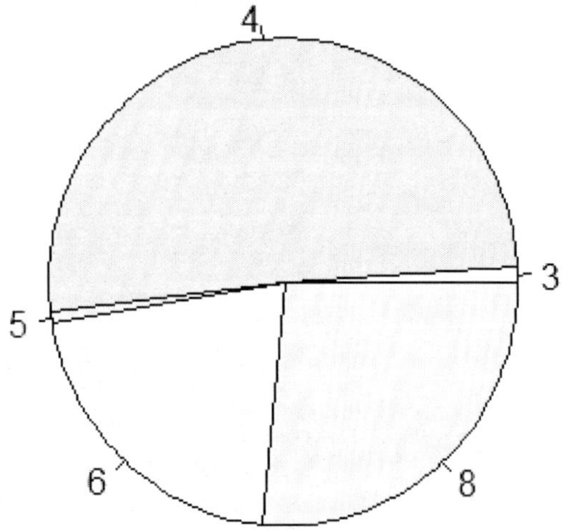

Figure 6-1

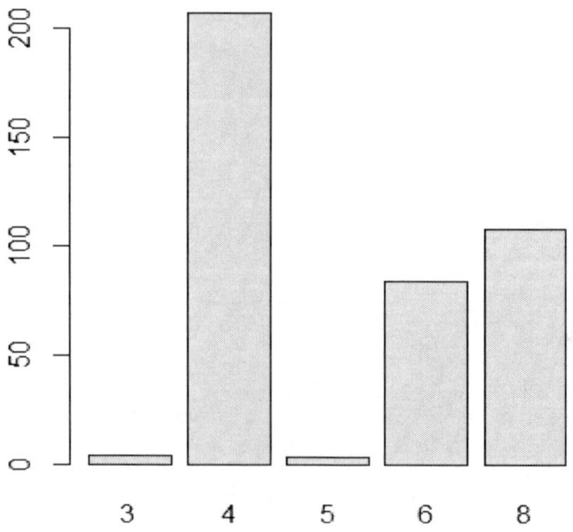

Figure 6-2

We can also build the contingency table of the *origin* and *cilinder* variables using the *table* and *xtabs* commands.

```
> table(origin,cylinder)
        cylinder
origin     3    4    5    6    8
      1     0 72    0   74  108
      2     0 66    3    4    0
      3     4 69    0    6    0
```

```
> xtabs(~origin+cylinder)
      cylinder
origin    3  4   5   6   8
     1    0 72   0  74 108
     2    0 66   3   4   0
     3    4 69   0   6   0
```

Exercise 1. The grades in Algebra and Calculus of 7 students are the following:

Álgebra	1	2	3	4	5	6	7
Cálculo	3	4	1	2	7	5	6

To determine the degree of association or agreement of the students' grades in Algebra and Calculus.

We calculate the correlation coefficient and the covariance of the two variables (declared as vectors).

```
> algebra=c(1,2,3,4,5,6,7)
> calculation=c(3,4,1,2,7,5,6)
> cor(algebra,calculus)
[1] 0.6071429
> cov(algebra,calculus)
[1] 2.833333
```

It is observed that they are positively correlated with a correlation coefficient of 0.6 which can be considered significant. In addition, the covariance is positive, which indicates that both variables vary in the same direction

Exercise 2. Ten authors compete for a novel prize, being classified by three different judges X, Y and Z in the following orders:

J1	1	6	5	10	3	2	8	7	9	4
J2	2	4	3	9	7	10	6	5	1	8
J3	5	3	9	6	4	2	8	7	10	1

Analyze which pair of judges are closest to your preferences.

When dealing with sorting scores it will be necessary to use the Sperman correlation to the variable matrix.

```
> J1=c(1,6,5,10,3,2,8,7,9,4)
> J2=c(2,4,3,9,7,10,6,5,1,8)
> J3=c(5,3,9,6,4,2,8,7,10,1)
> M=cbind(J1,J2,J3)

> M
      J1 J2 J3
 [1,] 1   2 5
 [2,] 6   4 3
 [3,] 5   3 9
 [4,] 10  9 6
 [5,] 3   7 4
 [6,] 2  10 2
 [7,] 8   6 8
 [8,] 7   5 7
 [9,] 9   1 10
[10,] 4   8 1

> cor(M,method="spearman")
 J1            J2          J3
J1 1.00000000 -0.07878788 0.5878788
J2 -0.07878788 1.00000000 -0.6121212
J3 0.58787879 -0.61212121 1.0000000
```

We observe that the closest judges in their preferences are 2 and 3 because they have the highest possible correlation in Spearman's correlation matrix.

Exercise 3. We consider the variables X, Y ,Z whose values are the following:

X	*Y*	*Z*
2	4	2
3	5	4
6	10	6
8	11	7
10	15	10

Find the correlation matrix by deducing the degree of dependence of the variables. Find the covariance matrix and interpret the result.

We begin by defining an array with the variables of the exercise, to then apply the commands that calculate the correlation matrix and the covariance matrix.

```
> X=c(2,3,6,8,10)
```

```
> Y=c(4,5,10,11,15)
> Z=c(2,4,6,7,10)
> matrix=cbind(X,Y,Z)
> matrix
      X  Y Z
[1,] 2  4 2
[2,] 3  5 4
[3,] 6 10 6
[4,] 8 11 7
[5,]10 15 10
> cor(matrix)
          X         Y         Z
X 1.0000000 0.9899319 0.9802123
Y 0.9899319 1.0000000 0.9830213
Z 0.9802123 0.9830213 1.0000000
```

In the previous output, it is observed that the correlation coefficient between variables *X* and *Y* is 0.9899, between *X* and *Z* is 0.9802, and between *Y* and *Z* is 0.9830, which indicates the strong positive linear dependence between each pair of variables. The correlation matrix has the following form:

$$\begin{pmatrix} r_{xx} & r_{xy} & r_{xz} \\ r_{xy} & r_{yy} & r_{yz} \\ r_{xz} & r_{yz} & r_{zz} \end{pmatrix} = \begin{pmatrix} 1 & 0{,}9899 & 0{,}9802 \\ 0{,}9899 & 1 & 0{,}9830 \\ 0{,}9802 & 0{,}9830 & 1 \end{pmatrix}$$

To find the covariance matrix we use the following syntax:

```
> cov(matrix)
        X     Y     Z
X    11.20 15.0  9.95
Y    15.00 20.5 13.50
Z     9.95 13.5 9.20
```

In view of the results, we can say that since all covariances are positive, the pairs of variables vary in the same direction around their means.

On the other hand, since the covariance between *X* and *Y is* 12, between *X* and *Z is* 7.96 and between *Y* and Z is 10.8 and in addition the variances of X, Y and Z are respectively 8.96, 16.4 and 7.36 the covariance matrix has the following form:

$$\begin{pmatrix} \sigma_x^2 & \sigma_{xy} & \sigma_{xz} \\ \sigma_{xy} & \sigma_y^2 & \sigma_{yz} \\ \sigma_{xz} & \sigma_{yz} & \sigma_z^2 \end{pmatrix} = \begin{pmatrix} 8{,}96 & 12 & 7{,}96 \\ 12 & 16{,}4 & 10{,}8 \\ 7{,}96 & 10{,}8 & 7{,}36 \end{pmatrix}$$

Exercise 4. We consider the file coches.sav and we want to build the multiple contingency table between the categorical variables origin, cylinder and derivative.

To build the contingency table of the *origin, cylinder* and *derivative* variables we use directly the *xtabs* command.

```
> xtabs(~origin+cylinder)
 Cylinder

origin     3    4    5    6   8
      1     0   72    0    74 108
      2     0   66    3    4    0
      3     4   69    0    6    0
> xtabs(~origin+cylinder+derivative)

, , derivative = 0



       cylinder
origin      3    4    5    6    8
      1     0    0    0    0  108
      2     0    0    0    0    0
      3     0    0    0    0    0

, , derivative = 1

       cylinder
origin      3    4    5    6    8
      1     0   72    0   74    0
      2     0   66    3    4    0
      3     4   69    0    6    0
```

Exercise 5. Build two functions that calculate the variance and quasi-variance of two variables Calculate the variance and quasi-variance of the first 10 natural numbers

We define the variance of a variable x as

$$\sigma_x^2 = \frac{1}{N}\sum_{i=1}^{n}(x_i - \bar{x})^2$$

Now we build a function in R for the variance of a variable x.

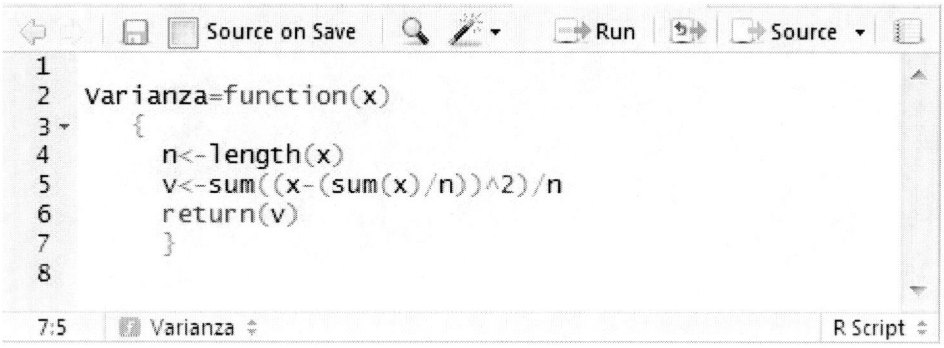

Next we calculate the variance of the first 10 natural numbers.

```
> x=c(1,2,3,4,5,6,7,8,9,10)
> Variance(x)
[1] 8.25
```

Now we build a function in R for the quasivariance of a variable *x*.

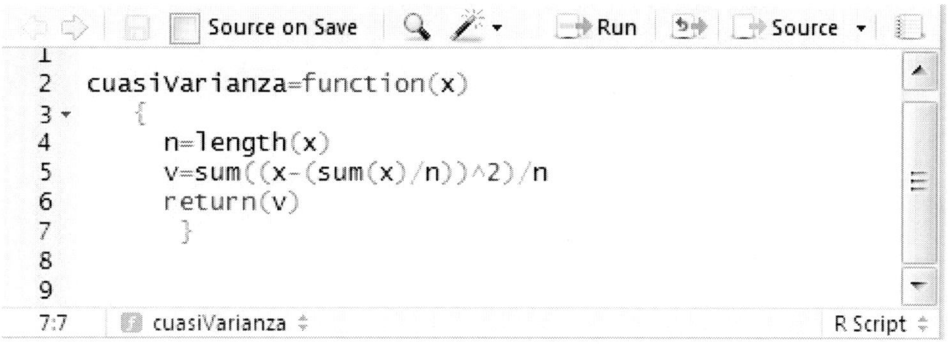

We then calculate the quasi-variance of the first 10 natural numbers.

```
> quasi-variance(x)
[1] 9.166667
```

Chapter 7

REGRESSION MODELS THROUGH R

7.1 MULTIPLE LINEAR REGRESSION

The objective of the multiple regression model is to explain the behaviour of an endogenous variable, explained or dependent, which we will designate as *Y*, using the information provided by the values taken by a set of explanatory variables, exogenous or independent, which we will designate as *X1, X2, ..., Xk*.

The linear model (econometric model) is given by the shape:

$$Y = \beta_0 + \beta_1 X_1 + \beta_2 X_2 + ... + \beta_k X_k + u$$

The coefficients (parameters) β_1, β_2, ..., β_k denote the magnitude of the effect that the explanatory variables (exogenous or independent) *X1, X2 , ..., Xk* have on the explained variable (endogenous or dependent) *Y*. The coefficient $_0$ is called the constant (or independent) term of the model. The term is called the εmodel error term.

We have a set of *T* observations for each of the endogenous and exogenous variables, one for each time period. Then, we can write the model of the form:

$$Y_t = \beta_0 + \beta_1 X_{1t} + \beta_2 X_{2t} + ... + \beta_k X_{kt} + u_t \quad t = 1,2,3, ...,T$$

The appearance (not necessary) of an independent term in the model can be interpreted as the presence of a first variable *X0* whose value is always 1.

The fundamental problem addressed is the following: assuming that the relationship between the variable *Y* and the set of variables *X1, X2, ... Xk* is as described in the model, and that there is a set of *T* observations in time for each of the variables, the endogenous and the exogenous ones, how can numerical values be assigned to the parameters $\beta_{0,}$, β_1, β_2, ... ,β_k based on the sample information? These values will be called parameter estimates.

Once the model parameter estimates are found, we can make predictions about the future behavior of the variable *Y*.

7.2 ASSUMPTIONS IN THE MULTIPLE REGRESSION MODEL

The basic assumptions that any linear model must meet can be classified into four major groups according to the components of the model itself. The first group of assumptions are those relating to the random disturbance *u*. The second group of assumptions are those relating to regressors. A third group of assumptions is made about the functional form. The last group of hypotheses are those concerning the parameter vector ($\beta_{0,}$, β_1, β_2, ... ,β_k).

7.2.1 Assumptions for error validation

We formulated the linear model under the following classical assumptions concerning the random disturbance *u*.

- The variable *u* (error term) is a random variable with zero hope and constant and diagonal covariance matrix (scalar matrix). That is, for every time instant *t,* the variable ε_t has a mean of zero and variance σ^2 not dependent on *t,* and in addition *Cov(ui, uj)* = *0* for all *i* and for all *j* time instants different from each other. The fact that the variance of *εt is* constant for all t (which does *not* depend on *t*) is called the *homoscedasticity* hypothesis and can also be expressed as V(*u*|*X1,X2, ..., Xk*) = σ^2 and V(*Y*|*X1,X2, ..., Xk*) =σ^2.

- The error term *u* is a random unobservable variable, which implies that the variable *Y is random,* since it depends on the random variable *u*.

- Also considered is the assumption of *normality of the* residues, consisting of the variables ε_t being normal for all *t. In* other words, the model's random disturbance vector has a multivariate normal distribution of zero mean E(***u***) = **0** and scalar covariance matrix E(***uu'***) = σ^2I. We can write ***u*** ***N***→**(0,**σ^2**I)**.

7.2.2 Hypotheses for the validation of regressors

The following assumptions regarding the regressors (independent or exogenous variables) in the model should be taken into account:

- The variables *X1, X2, ..., Xk* are linearly independent, i.e. there is no exact linear relationship between them. This hypothesis is called the *independence* hypothesis, and when it is not fulfilled, we say that the model has *multicollinearity*. It is equivalent to say that the regressor matrix has *k-rank*.

- The variables *X1, X2, ..., Xk* are deterministic (they are not random variables), since their value is constant from a sample taken over time, and furthermore they are not correlated with the error term *u, that* is, E(*u*|*X1,X2, ..., Xk*) = 0 (*exogeneity* hypothesis).

- The regressors have no observation or measurement errors.

7.2.3 Hypotheses for the parameters

- The parameter vector is a fixed vector.

This hypothesis ensures the stability over time of the estimates. The in-depth study of this hypothesis will lead us to the theory of cointegration, which is a topic that will be discussed later.

7.2.4 Assumptions for the model equation

- The relationship between *Y* and *X1, X2, ..., Xk* is indeed linear (*linearity* hypothesis).

- The absence of specification errors is also assumed, i.e. we assume that all *X* variables that are relevant for the explanation of the *Y* variable are included in the definition of the linear model.

7.3 OLS ESTIMATION (ORDINARY LEAST SQUARES)

Suppose we want to fit the following multiple linear regression model:

$$Y = \beta_0 + \beta_1 X_1 + \beta_2 X_2 + ... + \beta_k X_k + u$$

We have a set of *T* observations for each of the endogenous and exogenous variables (one for each moment in time). Then, we can write the model of the form:

$$Y_t = \beta_0 + \beta_1 X_{1t} + \beta_2 X_{2t} + ... + \beta_k X_{kt} + u_t \quad t = 1,2,3, ...,T$$

The MCO least squares criterion considers that the function that best fits the data is the one that minimizes the variance of the error u, which is equivalent to minimizing:

$$S(\beta_0, \beta_1, \cdots, \beta_k) = \sum_{t=1}^{T} u_t^2 = \sum_{t=1}^{T} (y_t - (\beta_0 + \beta_1 x_{1t} + \beta_2 x_{2t} + \cdots + \beta_k x_{kt}))^2$$

Deriving from parameters $_0$, $_{1,...,k}$, and equaling zero we have:

$$\left.\begin{array}{l}
\dfrac{\partial S}{\partial \beta_0} = 2\sum_{t=1}^{T} (y_t - (\beta_0 + \beta_1 x_{1t} + \beta_2 x_{2t} + \cdots + \phi_k x_{kt}))(-1) = 0 \\
\dfrac{\partial S}{\partial \beta_1} = 2\sum_{t=1}^{T} (y_t - (\beta_0 + \beta_1 x_{1t} + \beta_2 x_{2t} + \cdots + \beta_k x_{kt}))(-x_{1t}) = 0 \\
\vdots \\
\dfrac{\partial S}{\partial \beta_k} = 2\sum_{t=1}^{T} (y_t - (\beta_0 + \beta_1 x_{1t} + \beta_2 x_{2t} + \cdots + \beta_k x_{kt}))(-x_{kt}) = 0
\end{array}\right\} \Rightarrow$$

$$\Rightarrow \left\{\begin{array}{l}
\sum_{t=1}^{T} y_t = T\beta_0 + \beta_1 \sum_{t=1}^{T} x_{1t} + \cdots + \beta_k \sum_{t=1}^{T} x_{kt} \\
\sum_{t=1}^{T} y_t x_{1t} = \beta_0 \sum_{t=1}^{T} x_{1t} + \beta_1 \sum_{t=1}^{T} x_{1t}^2 + \cdots + \beta_k \sum_{t=1}^{T} x_{1t} x_{kt} \\
\vdots \\
\sum_{t=1}^{T} y_t x_{kt} = \beta_0 \sum_{t=1}^{T} x_{kt} + \beta_1 \sum_{t=1}^{T} x_{kt} x_{1t} + \cdots + \beta_k \sum_{t=1}^{T} x_{kt}^2
\end{array}\right.$$

These equations form a system called the system of normal equations, which can be solved for $_0$, $_1$, ..., $_k$ *by* any appropriate method to solve systems of linear equations. In this way the estimation of the model is obtained.

Alternatively, we can start from the matrix notation of the model $y = X\beta + u$, where *y is* the vector $T \times 1$, *X is a* matrix $T \times k$, *β is* a vector $k \times 1$ and *u* is a vector $T \times 1$ $\hat{y} = X\hat{\beta}$ $\hat{u} = y - \hat{y} = y - X\hat{\beta}$. If we designate with *S the* sum of the squares of the residues, we will have:

$$S = \hat{u}'\hat{u} = (\hat{u}_1, \hat{u}_2, \ldots, \hat{u}_T)\begin{bmatrix} \hat{u}_1 \\ \hat{u}_2 \\ \vdots \\ \hat{u}_T \end{bmatrix} = \sum_{t=1}^{T} \hat{u}_t^2$$

Which can also be written as:

$$S = (y - X\hat{\beta})'(y - X\hat{\beta}) = y'y - \hat{\beta}'X'y - y'X\hat{\beta} + \hat{\beta}'X'X\hat{\beta} = y'y - 2\hat{\beta}'X'y + \hat{\beta}'X'X\hat{\beta}$$

since ($\hat{\beta}'X'y = y'X\hat{\beta}$ it is the same climb in both expressions).

Applying the least-squares criterion is equivalent to minimizing the scale *S*. To minimize *S, the* first derivative of *S* is calculated with respect to the least-squares coefficient vector, $\hat{\beta}$, and it is obtained:

$$\frac{\delta S}{\delta \hat{\beta}} = -2X'y + 2X'X\hat{\beta}$$

By equating to zero, the system of normal equations is obtained:

$$X'X\hat{\beta} = X'y$$

To be able to solve the system with respect to $\hat{\beta}$ uniqueness, the range of the *X'X* matrix must be fulfilled to be equal to *k*. If this condition is met, both system members can be premultiplied by $[X'X]^{-1}$:

$$[X'X]^{-1}[X'X]\hat{\beta} = [X'X]^{-1}X'y \Rightarrow \hat{\beta} = [X'X]^{-1}X'y$$

We already have the expression of the minimum quadratic estimator vector:

$$\hat{\beta} = [X'X]^{-1}X'y$$

In addition, *S has* a minimum in $\hat{\beta}$, since the matrix of second derivatives, 2X'*X,* is defined as positive.

7.4 MAXIMUM LIKELIHOOD ESTIMATION

We want to estimate by maximum likelihood β and σ² of the model $y = X\beta + u$. To do this we will take as estimators those values that maximize the joint probability density for a sample of T observations, using the basic hypothesis e $u \rightarrow N(0, \sigma^2 I)$ $y \rightarrow N(X\beta, \sigma^2 I)$.

The probability density of *and* or the likelihood function, considering X and y fixed and β and $^{\sigma 2}$ variables:

$$L = \frac{1}{(2\pi\sigma^2)^{T/2}} e^{-(1/2\sigma^2)(y-X\beta)'(y-X\beta)}$$

As the maximum for *L is* reached at the same point as for *lnL,* because it is the monotonic logarithm function, we can, for maximization purposes, work with *lnL* instead of *L*. So:

$$\ln L = -\frac{T\ln(2\pi)}{2} - \frac{T\ln\sigma^2}{2} - \frac{1}{2\sigma^2}(y - X\beta)'(y - X\beta).$$

In order to maximise the use of the *L,* we refer to β and σ2:

$$\frac{\delta \ln L}{\delta\beta} = -\frac{1}{2\sigma^2}(-2X'y + 2X'y + 2X'X\beta),$$

$$\frac{\delta \ln L}{\delta\sigma^2} = -\frac{T}{2\sigma^2} + \frac{(y - X\beta)'(y - X\beta)}{2\sigma^4}.$$

By equalizing the partial derivatives to zero, the maximum plausible estimator of β, which we will denote by $\tilde{\beta}$, fulfills that:

$$X'X\tilde{\beta} = X'y$$

If we assume that $X'X$ is reversible, we have

$$\tilde{\beta} = [X'X]^{-1} X'y$$

So we have that the maximum-verifiable estimator of β, under the basic hypotheses, coincides with the minimum-square estimator. From the partial derivative with respect to σ^2, we can deduce that the maximum plausible estimator of σ^2 is $\frac{\hat{u}'\hat{u}}{T}$.

7.5 VALIDATION IN THE OLS MODEL

We already know that the multiple regression linear model with time series can be written in the form:

$$Y_t = \beta_0 + \beta_1 X_{1t} + \beta_2 X_{2t} + ... + \beta_k X_{kt} + u_t \quad t = 1,2,3, ...,T$$

The above expression can be represented in matrix form as follows:

$$\begin{pmatrix} Y_1 \\ Y_2 \\ \vdots \\ Y_t \end{pmatrix} = \begin{pmatrix} 1\, X_{11}\; X_{21} \cdots X_{k1} \\ 1\, X_{12}\; X_{22} \cdots X_{k2} \\ \cdots\cdots\cdots\cdots\cdots\cdots \\ 1\, X_{1T}\; X_{2T} \cdots X_{kT} \end{pmatrix} \begin{pmatrix} \beta_0 \\ \beta_1 \\ \vdots \\ \beta_k \end{pmatrix} + \begin{pmatrix} u_1 \\ u_2 \\ \vdots \\ u_t \end{pmatrix}$$

In short, we can put: $Y = X + u$

The first objective of econometric analysis is to obtain estimates, i.e. numerical values of the coefficients β_0, β_1, β_2, ..., β_k as a function of the sample information. These estimates can also be by intervals, that is, we can calculate confidence intervals for the parameters.

Let us assume that we already have a vector of estimates of the coefficients. We could write:

$$\hat{Y} = X\hat{B} = \hat{\beta}_0 + \hat{\beta}_1 X_1 + \hat{\beta}_2 X_2 + ... + \hat{\beta}_k X_k$$

$$\hat{Y}_t = \hat{\beta}_0 + \hat{\beta}_1 X_{1t} + \hat{\beta}_2 X_{2t} + ... + \hat{\beta}_k X_{kt} \quad t = 1,2,3,... ,T$$

Residues are, by definition, the differences between the true values of the variable *Yt* and the estimated values for *Yt*. That is, $\hat{u}_t = Y_t - \hat{Y}_t$ for all t.

From this we deduce that $Y = \hat{Y} + \hat{u} = X\hat{B} + \hat{u}$, so the original model is $Y = X\beta + u$, and the estimated model will be $Y = X\hat{B} + \hat{u}$. The parameter estimates can be calculated by the method of least squares, which consists of minimizing the sum of the squares of the *residuals,* also called the *residual sum* (*SR*), whose expression is as follows:

$$SR = \hat{u}'\hat{u} = \sum_{t=1}^{T} \hat{u}_t^{\,2} = \sum_{t=1}^{T} (Y_t - \hat{Y}_t)^2$$

We have already seen that the value of the *parameter estimates (by minimum squares or by maximum likelihood) is* given by the expression $\hat{B}=(X'X)^{-1}X'Y$. These estimates are unbiased, because $E(\hat{B})=B$. The covariance matrix $\hat{B}$ is $(X'X)^{-1}$:

$$\hat{\beta}=[X'X]^{-1}X'[X\beta+u]=[X'X]^{-1}X'[X\beta+u]=\beta+[X'X]^{-1}X'u$$

Applying hope and taking into account that β is a fixed vector and X *is* a fixed matrix, you have

$$E(\hat{\beta})=\beta+[X'X]^{-1}X'E(u)=\beta$$

The variance-covariance matrix of the minimum-square estimator vector is given by

$$E[\hat{\beta}-(\hat{\beta})][\hat{\beta}-(\hat{\beta})]'=E\left[(\hat{\beta}-\beta)(\hat{\beta}-\beta)'\right]=E\left[[X'X]^{-1}X'uu'X[X'X]^{-1}\right]=$$
$$=[X'X]^{-1}X'E[uu']X[X'X]^{-1}=[X'X]^{-1}X'[\sigma^2 I]X[X'X]^{-1}=\sigma^2[X'X]^{-1}$$

The variance of $\hat{\beta}_i$ will then be given by $\sigma^2 a_{ii}$, where *aii is* the *ii-th* element of $[X'X]^{-1}$. Similarly, the covariance between and will be $\hat{\beta}_i$ $\hat{\beta}_j$ $\sigma^2 a_{ij}$, where *aij is the i-th* element of $[X'X]^{-1}$.

From the previous results we deduce that the estimator of $\hat{\beta}_i$ any of the coefficients $\beta_{i,}$ has as mathematical hope $_{i,}$ and as standard deviation the value 2*aii* where *aii is* the *i-th* element in the main diagonal of the matrix σ^2 $(X'X)^{-1}$.

Then, under the hypothesis of normality of the residues, the statistician

$$N_i=\frac{\hat{\beta}_i-\beta_i}{\sigma\sqrt{a_{ii}}}$$

follows a normal distribution (0.1).

The estimator (maximum plausible and minimum square) of σ^2 is $\frac{\hat{u}'\hat{u}}{T}$, but this estimator is not unbiased. An *unbiased error variance estimator* is

$$\hat{\sigma}^2 = \frac{\hat{u}'\hat{u}}{T-k-1}$$

On the other hand, it is also shown that the statistic *G=u'u/2* follows a Chi-square distribution with $T - k - 1$ degrees of freedom, which will allow us to calculate *confidence intervals and hypothesis contrasts for and to its square.*

The distributions of the *Ni* and *G statistics* lead us to the conclusion that the $N_i / [G/(T-k-1)]^{1/2}$ statistic is a *t* Student with *T– k –1* degrees of freedom, which allows us to state that the statistic:

$$T_i = \frac{\hat{\beta}_i - \beta_i}{\hat{\sigma}\sqrt{a_{ii}}}$$

follows a Student *t-distribution* with $T-k-1$ degrees of freedom, which will allow us to *find confidence intervals and hypothesis contrasts for the model's* β_i *parameters.* It will be possible to contrast the null hypothesis *Ho* that $\beta_i = 0$ for each *i=1.2, ...,T in the* usual way using the *Ti* statistic; that is, if *Ti0* is the value of *Ti* when $\beta_i = 0$, the hypothesis *Ho* will be accepted at the level α when $T_{i0} \leq t_{\alpha/2,T-k-1}$. The confidence interval for β_i at the level αwill be given by $\beta_i \pm t_{\alpha/2,T-k-1}\ \hat{\sigma}\sqrt{a_{ii}}$ ($t_{\alpha/2,T-k-1}$ is the value of the abscissa of a Student *t with* $T-k-1$ degrees of freedom, which leaves $\alpha/2$ area to its right).

We now introduce the following concepts:

- $Total\ sum\ ST = \sum_{i=1}^{T}(Y_t - \bar{Y})^2 = Y'Y - T\bar{Y}^2$.
- $Eexplained\ \ sum\ SE = \sum_{i=1}^{T}(\hat{Y}_t - \bar{Y})^2 = \hat{Y}'Y - T\bar{Y}^2$.
- $Residual\ sum\ (SR = \sum_{i=1}^{T}(Y_t - \hat{Y})^2 = \hat{u}'\hat{u}$ concept already known).

The *sum total* is the sample variance of the endogenous variable (except for the sample size factor), and is therefore a measure of the size of the fluctuations experienced by that variable around its mean value. The fundamental purpose of any econometric model is to try to explain these fluctuations.

The *sum explained* is the degree of fluctuation of the variable around the average of *Y*. Therefore, the explained *sum is the* level of fluctuation of the variable *Yt* that the model is able to explain. It is the variation explained by the regressors.

The *residual sum*, already introduced previously, is an indicator of the level of error of the model in its attempt to explain the temporal evolution of the variable *Yt*.

We already know that:

$$SR = \sum_{i=1}^{T}(Y_t - \hat{Y})^2 = \hat{u}'\hat{u} = (Y - X\hat{B})'(Y - X\hat{B}) = Y'Y - \hat{B}'X'Y = Y'Y - \hat{Y}'Y$$

Then we can write equality, $Y'Y = \hat{Y}'Y + \hat{u}'\hat{u}$ and if we subtract $T\bar{Y}^2$ from this equality, we have to:

$(Y'Y - T\bar{Y}^2) = (\hat{Y}'Y - T\bar{Y}^2) + \hat{u}'\hat{u}$ that is, *ST* = *SE* + *SR*.

Then you have: total sum = explained sum + residual sum

These three terms are called the *sum of squares.*

Each sum of squares divided by their degrees of freedom is called a mean square. Under the hypothesis of normality of residues, *SE* is distributed according to a Chi-square with *k* degrees of freedom, *SR according to a* Chi-square with $T–k–1$ degrees of freedom, and *ST* according to a *Chi-square* with $n–1$ degrees of freedom. Therefore the Chi-square *explained by the model* will be *CM(E)* = *SE/k,* and the *residual Chi-square* will be *CM(R)* = *SR/(T – k – 1)*.

The *determination coefficient* (*R2*) is defined as a descriptive measure of the overall fit of the model whose value is the ratio of the explained variability (or *explained sum) to the* total variability (or *total sum*), i.e. *R2* = *SE/ST* = 1 – *SR/ST.*

A model will be all the better the higher the *R2,* although this statement is not too severe, since this coefficient depends very much on new variables introduced into the model, even if these do not worsen the quality of the regression.

This problem is solved by replacing this coefficient with the corrected coefficient of determination, which for large samples will no longer depend on the number of variables in the model.

The *multiple correlation coefficient is* defined as the square root of the determination coefficient, and its value is *R*.

The coefficient of determination corrected for degrees of freedom (adjusted coefficient of determination) is defined as the value:

$$\overline{R}^2 = 1 - (1 - R^2)\frac{T-1}{T-k-1}$$

It is observed that when $T \rightarrow \infty$, that is, for large samples, $(T-1)/(T-k-1)$ $1 \rightarrow$ and does not depend on k, which is the number of variables in the model. In addition, $T \rightarrow \infty \Rightarrow \overline{R}^2 \rightarrow R^2$

We can now consider a $\overline{R}^2$ as a good measure of the quality of the regression. The higher the corrected coefficient of determination, the better the model will be $\overline{R}^2$.

From the *SE* and *SR* distributions, it follows that the statistician

$$F = \frac{SE/k}{SR/(T-k-1)}$$ tenth una distribución $F(k, T-k-1)$ de Fisher-Snedecor.

But as $1 - R2 = SR/ST$, we can put *F of the* form:

$$F(k, T-k-1) = \frac{R^2}{(1-R^2)}\frac{(T-k-1)}{k}$$

Therefore, the *F* statistic will allow us to make *comparisons on the correlation coefficient*. For the case of simple regression ($k = 1$) we have an $F(1, T-2)$ that is equivalent to a *t of* Student with $T-2$ degrees of freedom.

The statistician follows an $\frac{(\hat{B}-B)' X' X(\hat{B}-B)}{k\hat{\sigma}^2}$ $F(k, T\,k\,1$ ––) distribution.

This statistic will allow us to find *regions of confidence at a level of significance* α *for parameter set* β_i *of the model*. This statistic will also allow us to contrast the null hypothesis $\beta_1 = \beta_2 = \ldots = \beta_k = 0$.

The Analysis of Variance table will look like this:

Source of Variation	*Sum of squares*	*Degrees of freedom*	*Squares media*	*F*
Model *residual*	*N* *SR*	*k* *T-k-1*	*CM(E)=SE/k* *CM(R)=SR/(T-k-1)*	*CM(E)* *CM(R)*
Total	*ST*	*T-1*		

The most general statistician:

$$T = \frac{(D\hat{B} - DB)'[D(X'X)^{-1}D']^{-1}(D\hat{B} - DB)}{k\hat{\sigma}^2}$$

also follows an $F(k, T-k-1)$ distribution for a suitable matrix D.

This statistic will allow more general contrasts of various kinds of hypotheses and build regions of confidence for the model parameters and for the predictions. To do this, it is sufficient to take the appropriate forms of the D matrix.

To contrast the hypothesis $\beta_1 = \beta_1^*$, $\beta_2 = \beta_2^*, \ldots, \beta_k = \beta_k^*$, we take:

$$D = \begin{pmatrix} 1 & 0 & \cdots & 0 \\ 0 & 1 & \cdots & 0 \\ \vdots & \vdots & \ddots & \vdots \\ 0 & 0 & \cdots & 1 \end{pmatrix} = I_{kxk} \Leftrightarrow DB = \begin{pmatrix} 1 & 0 & \cdots & 0 \\ 0 & 1 & \cdots & 0 \\ \vdots & \vdots & \ddots & \vdots \\ 0 & 0 & \cdots & 1 \end{pmatrix} \underbrace{\begin{pmatrix} \beta_1 \\ \beta_2 \\ \vdots \\ \beta_k \end{pmatrix}}_{B} = \underbrace{\begin{pmatrix} \beta_1^* \\ \beta_2^* \\ \vdots \\ \beta_k^* \end{pmatrix}}_{B^*} \Leftrightarrow$$

$$\Leftrightarrow T = \frac{(\hat{B} - B^*)'X'X(\hat{B} - B^*)}{k\hat{\sigma}^2} \rightarrow F_{k,T-k-1}$$

To contrast a subset of parameters $\beta_{r+1} = \beta_{r+1}^*$, $\beta_{r+2} = \beta_{r+2}^*, \ldots, \beta_{r+k} = \beta_{r+k}^*$, we take:

$$D = \left(\begin{matrix} 0 & 0 & \cdots & 0 \\ 0 & 0 & \cdots & 0 \\ \vdots & \vdots & \ddots & \vdots \\ 0 & 0 & \cdots & 0 \end{matrix} \right._{r} \left. \begin{matrix} 1 & 0 & \cdots & 0 \\ 0 & 1 & \cdots & 0 \\ \vdots & \vdots & \ddots & \vdots \\ 0 & 0 & \cdots & 1 \end{matrix} \right)_{k-r} = I_{kxk} \Leftrightarrow$$

$$\Leftrightarrow T = \frac{(D\hat{B} - B^*)'[D[X'X]^{-1}D']^{-1}(D\hat{B} - B^*)}{(k-r)\hat{\sigma}^2} \rightarrow F_{k-r,T-k-1}$$

It is also possible to contrast a set of linear restrictions on parameters, which can be written in general as follows:

$$\begin{aligned} a_{11}\beta_1 + a_{12}\beta_2 + \cdots + a_{1k}\beta_k &= \beta_1^* \\ a_{21}\beta_1 + a_{22}\beta_2 + \cdots + a_{2k}\beta_k &= \beta_2^* \\ &\vdots \\ a_{r1}\beta_1 + a_{r2}\beta_2 + \cdots + a_{rk}\beta_k &= \beta_r^* \end{aligned}$$

where:

$$D = \begin{pmatrix} a_{11} & a_{12} & \cdots & a_{1k} \\ a_{21} & a_{22} & \cdots & a_{2k} \\ \vdots & \vdots & \ddots & \vdots \\ a_{r1} & a_{r2} & \cdots & a_{rk} \end{pmatrix} \text{ y } T = \frac{(D\hat{B} - DB)'[D(X'X)^{-1}D'(D\hat{B} - DB)}{r\hat{\sigma}^2} \to F_{r,T-k-1}$$

The Gaussian-Markov theorem asserts that in the context of the linear regression model (under the typical assumptions) the MCO estimators of the parameters β_0, β_1, β_2, ..., β_k *are the* ones with the lowest variance between the linear and unbiased estimators. In addition, the MCO estimators $\hat{\beta}_0, \hat{\beta}_1, \hat{\beta}_2, \ldots, \hat{\beta}_k$ are consistent estimators of β_0, β_1, β_2, ..., β_k

$$p\lim \hat{\beta}_j = \beta_j \quad j = 0,1,\ldots,k$$

In other words, MCO estimators $\hat{\beta}_0, \hat{\beta}_1, \hat{\beta}_2, \ldots, \hat{\beta}_k$ are consistent estimators of the parameters β_0, β_1, β_2, ..., β_k because their bias tends to zero when the sample size tends to infinity.

7.6 PREDICTION IN THE OLS MODEL

A prediction can be defined as an attempt to anticipate the future. In the temporal context, and in the case of quantitative procedures, we can speak of two types of predictions: conditional and unconditional. ***Conditional predictions*** are those made by means of causal models. For example, in a time series regression model that relates two variables, one dependent, *Y,* and another independent, *X,* the predictions of *Y are* conditioned to *X, that* is, *Y* is predicted given *X.*

Unconditional predictions are those made through self-projective methods. These methods can be based on two alternative approaches: the deterministic, or classical, and the stochastic, or modern (based on the methodology of Box and Jenkins). The *deterministic approach* is more appropriate when a limited number of observations are available, while the stochastic *approach is* more appropriate when the series are larger.

For each type of prediction (short, medium and long term), there are certain more suitable methods. For example, trend analysis is a method for making long-term predictions, econometric models are suitable for making short- and medium-term predictions, and self-projective methods are more suitable for making short-term predictions.

As we have seen, one of the purposes of econometric model analysis is to make conditional predictions for the dependent variable. If we estimate the model *Y=XB* and we obtain the estimated model, $\hat{Y} = X\hat{B}$ we have that the value $\hat{Y}_0 = X_0\hat{B}$ is an optimal unbiased linear estimator of the forecast of *Y,* for a given value *X0* of *X.*

7.6.1 Point and average predictions and confidence intervals

You can predict the mean, *E*(*Y0*), or the point value, *Y0*. The prediction errors will be quantified by the variances of the predictors. The variance for the mean prediction is $X0(X'X)^{-1}X0'$.

A *confidence interval for the prediction in mean to* level will be α given by the expression $\hat{Y}_0 \pm t_{T-k-1}(\alpha/2)\hat{\sigma}\sqrt{X_0(X'X)^{-1}X_0'}$, where $t_{(T-k-1)}(\alpha/2)$ is the value of the Student *t* with $T-k-1$ degrees of freedom at the point **.**

The variance for the point prediction is $\sigma^2\ (X_0(X'X)^{-1}\ X_0'+1)$, and a *confidence interval for the point prediction* will be given by the expression defined as $\hat{Y}_0 \pm t_{T-k-1}(\alpha/2)\hat{\sigma}\sqrt{1+X_0(X'X)^{-1}X_0'}$, where $t_{(T-k-1)}(\alpha/2)$ is the value of the Student *t* with $T-k-1$ degrees of freedom at the point **.**

7.6.2 Indicators of the Predictive Capacity of a model

Several alternative statisticians are used to evaluate the predictive ability of the model with Eviews. Being *n* the prediction horizon, the most common statisticians for the assessment of predictive ability are the following:

Root Mean Squared Error:

$$RECM = \sqrt{\frac{\sum_{i=1^{\circ}}^{n}(\hat{Y}_i - Y_i)^2}{n}}$$

Mean Absolute Error

$$EAM = \frac{\sum_{i=1^{\circ}}^{n} |\hat{Y}_i - Y_i|}{n}$$

Mean Abs. Percent Error:

$$EAMP = \frac{1}{n}\sum_{i=1^{\circ}}^{n}\left|\frac{\hat{Y}_i - Y_i}{Y_i}\right|$$

Theil Inequality Coefficient

$$CDT = \frac{\sqrt{\frac{\sum_{i=1^{\circ}}^{n}(\hat{Y}_i - Y_i)^2}{n}}}{\sqrt{\frac{\sum_{i=1^{\circ}}^{n}\hat{Y}_i^2}{n}} + \sqrt{\frac{\sum_{i=1^{\circ}}^{n}Y_i^2}{n}}}$$

Bias Proportion:

$$\frac{(\overline{\hat{Y}} - Y)^2}{\sum_{i=1^{\circ}}^{n}(\hat{Y}_i - Y_i)^2 / n}$$

Variance Proportion

$$\frac{(S_{\hat{Y}} - S_Y)^2}{\sum_{i=1^{\circ}}^{n}(\hat{Y}_i - Y_i)^2 / n}$$

Covariance Proportion

$$\frac{2(1-r)S_{\hat{Y}}S_Y)}{\sum_{i=1^{\circ}}^{n}(\hat{Y}_i - Y_i)^2 / n}$$

The closer the values of the first four statisticians are to zero, the better the predictive capacity of the model, allowing a model to be compared with alternative ones. The three ratios vary between zero and one and should also be small.

7.7 OPTIMAL REGRESSION MODELS

When a multiple regression model is fitted we can have problems of different kinds. In this situation there are criteria that allow us to choose the best model for a given variable and data set.

Apart from the *forward regression criterion,* which includes variables in the model until the ideal fit is obtained, the *backward regression criterion* that starts by including all variables in the model and eliminates the appropriate ones until an optimal fit is obtained free of problems, and the step-by-step *selection criterion,* there are other model selection criteria based on statistics such as Mallows' *R2* and *Cp,* which allow to obtain well-fitted models very useful in exploratory fits.

In addition, the AIC statistics of AKAIKE and SC of Schwarz allow to *select the adjusted model with better explanatory capacity* that one that presents smaller value for these statistics. The definition of each one of them is:

$$AIC = -\frac{2l}{T} + \frac{2(K+1)}{T}$$

$$SC = -\frac{2l}{T} + \frac{(K+1)\log(T)}{T}$$

$$l = -\frac{T}{2}(1 + \log(2\pi) + \log\frac{e'e}{T})$$

Where K is the number of independent model variables (not including the constant), T is the sample size or number of observations available for model estimation and e is the model error.

7.8 ERROR DIAGNOSIS

Once the model of regression with time series has been constructed, we will have to contrast, among others, the hypotheses of linearity, normality, homoscedasticity, non-correlation and independence, matter that will be extended in successive chapters of this book. The residues will present a first information about these hypotheses.

If the frequency histogram of the residues does not match that of a normal one, outliers may exist. By removing the pairs (*Xi Yi*) that produce the outliers, *normalization of the* residue can be achieved.

If we plot the values of t against the values of $\hat{u}_t$ (that $\hat{u}_t$ is, if we make the graph whose points are the pairs $(t, \hat{u}_t)$) and we detect an increasing or decreasing trend in the network, there may be *autocorrelation or serial correlation*.

If we graph the values $\hat{Y}_t$ against the values of $\hat{u}_t$, that is, if we make the graph whose points are the pairs ($\hat{Y}_t$,$\hat{u}_t$) and we detect a trend of any kind in the network, there can be autocorrelation, since there will be a correlation between the residuals. Also in this case there can be *heteroscedasticity,* or also lack of *linearity.*

If we plot the values $\hat{Y}_t$ against the values of $\hat{u}_t^{2}$, that is, if we make the graph whose points are the pairs ($\hat{Y}_t$,$\hat{u}_t^2$) and we detect a trend of any kind in the network, there may be heteroscedasticity.

If we plot the values of *Xt* against the values of $\hat{u}_t$, that is, if we make the graph whose points are the pairs (*Xt*, $\hat{u}_t$) and we detect an increasing or decreasing trend in the network, there may be an autocorrelation, since the residues will not be uncorrelated with the explanatory variables. There may also be heteroscedasticity, or lack of linearity.

If we plot the values of *Xt* against the values of $\hat{u}_t^2$, that is, if we make the graph whose points are the pairs (*Xt*, $\hat{u}_t^2$) and detect any trend in the graph, there may be heteroscedasticity or nonlinearity (there will be a relationship between the variance of the error term and the explanatory variables).

These analyses can also be performed using standardized residues or studied residues, which are usually more effective in detecting deficiencies in the model.

The studied residues, whose distribution is a t Student with T–k–2 degrees of freedom, are also used to detect outliers in the residues (analysis of the normality of the residues or of the poor specification of the model).

A very important contrast for detecting autocorrelation is the Durbin-Watson contrast, according to which the statistician

$$d = \frac{\sum_{t=2}^{T} (\hat{u}_t - \hat{u}_{t-1})^2}{\sum_{t=1}^{T} \hat{u}_t^{\,2}}$$

allows us to adopt the not too rigorous rule that if *d is* worth 0 there is positive perfect autocorrelation; if *d is* close to 2 there is no autocorrelation, and if *d is* close to 4 there is negative perfect autocorrelation. However, *d is* tabulated, and depending on the range in which its value falls, the autocorrelation hypothesis is accepted or rejected.

If the $X'X$ matrix has a zero determinant, its inverse cannot be calculated (nor can parameter estimates be made), in which case there is multicollinearity. This problem is usually solved by replacing the variables of the model by a large set of its main components to do the regression again, but this time with the main components as independent variables of the model. Multicollinearity can also be addressed by means of the correlation matrix of the independent variables, since highly correlated variables may be candidates for multicollinearity, a problem that can be solved by eliminating one of the strongly correlated variables, as long as that variable is not relevant.

7.9 RESTRICTIONS IN THE MULTIPLE REGRESSION MODEL

Let us suppose that certain hypotheses are contrasted in a model that can later be fixed a priori. In order to estimate a new similar model, it may be interesting to do so by already fixing the restrictions derived from the previously contrasted certain hypothesis. The problem now will be to find the estimator $\hat{\beta}_R$ that minimizes the sum of squares of the residuals, in the same way that we did in the linear model of multiple regression, but this time imposing the restrictions $R\beta = r$, that is, unlike what was done there, this time it is a matter of solving an optimization problem subject to linear restrictions. The Lagrangian of such problem will be:

$$L = (y - X\beta)'(y - X\beta) - 2\lambda'(R\beta - r)$$

where λ is a vector of dimension $q \times 1$ of Lagrange multipliers (as many as constraints). By taking partial derivatives of this Lagrangian with respect to β', as well as with respect to λ, *we* have

$$\frac{\partial L}{\partial \beta'} = -2X'y + 2X'X\beta - 2R'\lambda$$

$$\frac{\partial L}{\partial \lambda} = -2(R\beta - r)$$

Equalizing these partial derivatives to zero and solving the system of $k + q$ equations thus obtained results:

$$X'X\beta - X'y - R'\lambda = 0$$

$$R\beta_R - r = 0$$

The solutions of this system of equations lead us to the Restricted Least Squares Estimator (RCM).

Premultiplying by $X'X\beta - X'y - R'\lambda = 0$ $R(X'X)^{-1}$ you have:

$$R\beta - R(X'X)^{-1}X'y - R(X'X)^{-1}R'\lambda = 0$$

Under the null hypothesis $R\beta = r$ *you* have:

$$\lambda = \left[R(X'X)^{-1}R'\right]^{-1}\left(r - R\hat{\beta}\right)$$

where $\hat{\beta}$ it is the usual MCO estimator, i.e. without imposing any of the restrictions contained in Ho. By substituting in λ **and** $X'X\beta - X'y - R'\lambda = 0$ calling $\hat{\beta}_R$ the vector of estimated parameters with restrictions, you have:

$$X'X\hat{\beta}_R - X'y - R\left[R(X'X)^{-1}R'\right]^{-1}\left(r - R\hat{\beta}\right) = 0$$

By multiplying the above expression to the left by $(X'X)^{-1}$ you have:

$$(X'X)^{-1}X'X\hat{\beta}_R - (X'X)^{-1}X'y - (X'X)^{-1}R\left[R(X'X)^{-1}R'\right]^{-1}\left(r - R\hat{\beta}\right) = 0$$

$$\hat{\beta}_R - \hat{\beta} - (X'X)^{-1} R\left[R(X'X)^{-1} R'\right]^{-1}\left(r - R\hat{\beta}\right) = 0$$

$$\hat{\beta}_R = \hat{\beta} + (X'X)^{-1} R'\left[R(X'X)^{-1} R'\right]^{-1}\left(r - R\hat{\beta}\right)$$

We have already calculated the $\hat{\beta}_R$ model's restricted least squares estimator (RCM). We can interpret this expression by saying that the restricted OLS estimator is an unrestricted estimator correction, with the size of the correction being the second term in the expression of $\hat{\beta}_R$. The farther the unrestricted estimator is from satisfying the constraints, the greater the correction.

The MCR estimator is unbiased only if the restrictions $R\beta = r$ under which it has been obtained are true. The MCR estimator differs from the MCO estimator only if the latter does not satisfy the restrictions in Ho (which will generally be the case). But, if it turns out that the MCO estimator satisfies exactly the restrictions whose validity is contrasted, then the MCR estimator coincides with the MCO estimator. The covariance matrix of the MCR estimator is always lower than the covariance matrix of the MCO estimator, even if the restrictions are not true. Although it may seem paradoxical, this result makes sense because, by imposing the restrictions, we limit the region of the parametric space in which we seek the MCO estimator, so we can estimate it with greater precision.

The restricted estimator covariance matrix is:

$$Var\left(\hat{\beta}_R\right) = \sigma_u^2\left[(X'X)^{-1} - (X'X)^{-1} R'\left[R(X'X)^{-1} R'\right]^{-1} R(X'X)^{-1}\right]$$

So:

$$Var\left(\hat{\beta}_R\right) - Var\left(\hat{\beta}\right) = \sigma_u^2 (X'X)^{-1} R'\left[R(X'X)^{-1} R'\right]^{-1} R(X'X)^{-1}$$

The latter expression is a positive defined matrix, which leads to the elements of the diagonal of the covariance matrix $\hat{\beta}_R$ being smaller than the corresponding elements of the covariance matrix of $\hat{\beta}$, for a given estimation of the parameter σ_u^2.

7.10 THE MULTIPLE LINEAR REGRESSION MODEL IN R

It is possible to obtain the following information regarding a model:

r=lm(y~x1+x2+x3+....)	*It performs linear least-squares regression with independent varibals xi and with dependent variable y*
summary(r)	*Displays detailed results of the fitted model*
coefficients(r)	*Lists the adjusted parameters of the model*
confint(r)	*Lists the 95% confidence intervals for the estimated model parameters*
fitted(r)	*List of predicted values in the adjusted model*
residuals(r)	*List the residues of the adjusted model*
anova(r)	*List the ANOVA table of the adjusted model*
vcov(r)	*Lists the covariance matrix of the estimated model parameters*
AIC(r)	*Return the AKAIKE statistic*
plot(r)	*Displays graphical diagnostics for the fitted model*
predict(r)	*Predicts values with the model fitted*

As an example let's consider the linear regression that makes the consumption of cars depend on their power, weight and engine.

```
> r1=lm(consumption~cv+weight+engine)
> r1

Call:
lm(formula = consumption ~ hp + weight + engine)

Coefficients:
(Intercept)          cv          engine          weight
 0.8762787    0.0412840     0.0051719       0.0003043

> summary(r1)

Call:
lm(formula = consumption ~ hp + weight + engine)

Residuals:
     Min       1Q Median       3Q      Max
-5.1803 -1.1196 -0.0541 0.8530 20.0023

Coefficients:
              Estimate   Std. Error  t value    Pr(>|t|)
(Intercept)    0.8762787 0.5427383    1.615      0.107
```

```
cv                 0.0412840 0.0057376    7.195      3.07e-12 ***
weight             0.0051719 0.0009563    5.408      1.10e-07 ***
engine            0.0003043  0.0001819    1.674      0.095 .
---
Meaning codes: 0 '***' 0.001 '**' 0.01 '*' 0.05 '.' 0.1 ' ' 1

Residual standard error: 1.938 on 402 degrees of freedom
Multiple R-squared: 0.7609,    Adjusted R-squared: 0.7591
F-statistic: 426.3 on 3 and 402 DF, p-value: < 2.2e-16
```

We observe that the adjusted R2 is quite high, the p-values of the Student T for the estimated parameters are low enough that they are all individually significant at least 90%. In addition, since the p-value of the F of Fisher is very small, the joint significance of the estimated parameters is also met.

To obtain 95% confidence intervals for the estimated model parameters we use the following syntax:

```
> confint(r1)
                        2.5 %         97.5 %
(Intercept)     -1.906812e-01 1.9432385272
cv               3.000464e-02 0.0525633722
weight           3.291844e-03 0.0070519317
engine          -5.316155e-05 0.0006618361
```

If we want to save the model residues in a variable we use the following syntax:

```
> residual=residuals(r1)
```

If we want to save the predicted values in a variable we use the following syntax:

```
> predicted=fitted(r1)
```

If I want to see the ANOVA table of the model we use the following syntax:

```
> anova(r1)
Analysis of Variance Table

Response: consumption
              Df  Sum Sq  Mean Sq    F value Pr(>F)
cv             1  4454.1  4454.1   1186.0506        <2e-16 ***
weight         1  338.7    338.7     90.1766        <2e-16 ***
engine         1   10.5     10.5      2.8008        0.095 .
```

```
Residuals     402 1509.7        3.8
---
Meaning codes: 0 '***' 0.001 '**' 0.01 '*' 0.05 '.' 0.1 ` ' 1
```

This table is usually used to test the joint significance of the model parameters through the p'value of Fisher's F (must be small).

To obtain the covariance matrix of the estimated parameters we use the following syntax:

```
> vcov(r1)
              (Intercept)            cv         engine         weight
(Intercept) 2.945649e-01 -9.591434e-04 -4.275433e-04 7.482507e-05
cv         -9.591434e-04  3.291957e-05 -7.696842e-07 -5.401682e-07
weight     -4.275433e-04 -7.696842e-07 9.145781e-07 -1.248696e-07
engine      7.482507e-05 -5.401682e-07 -1.248696e-07 3.306999e-08
```

As it has small values the model is well estimated.

We then evaluated the model's predictive ability with Akaike's AIC statistician.

```
> AIC(r1)
[1] 1695.379
```

One model predicts better than another when its AIC statistician is less.

Now we generate graphic diagnostics to perform the model diagnosis using the following syntax. We obtain the graph of the residues against the predicted values (as it is random there are no problems of heteroscedasticity or linearity). Figure 7-1 shows an outlier. Figure 7-2 shows the QQ graph of normality (it fits well to the diagonal of the first quadrant except the outlier). Figure 7-3 shows the predicted values against the studied residuals (their randomness reinforces linearity and homoscedasticity). Figure 7-4 shows the atypical residues being influential the observation 35 (it is out of the bands of the cook statistic). It would be best to eliminate this observation and re-estimate the model.

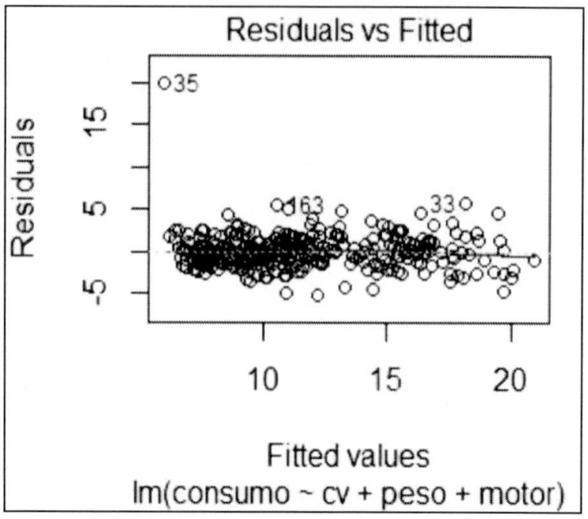

Figure 7-1

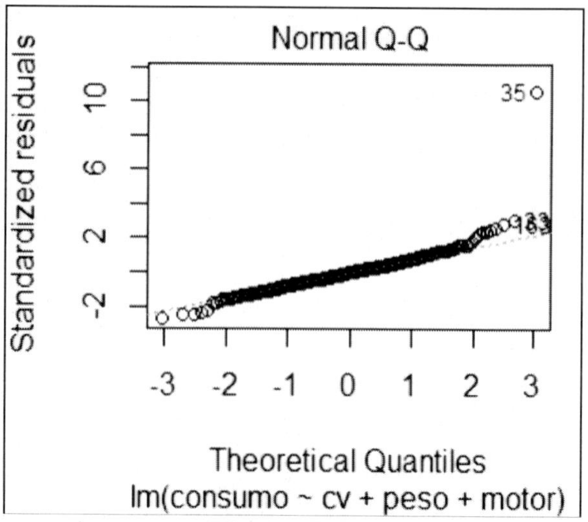

Figure 7-2

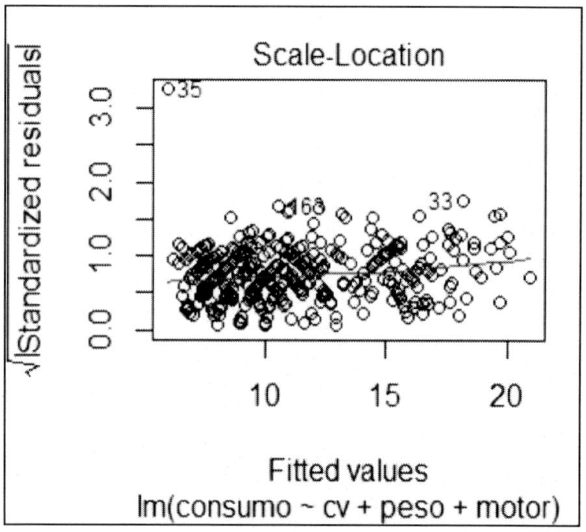

Figure 7-3

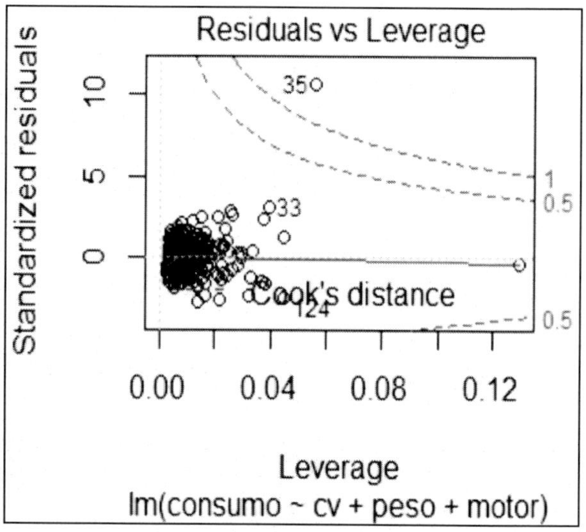

Figure 7-4

If we make the histogram of the residue we see that it fits a normal (Figure 7-5). The graph of the estimated residuals against the row number is random (Figure 7-6)

```
> hist(residues,xlim=c(-10,10))
```

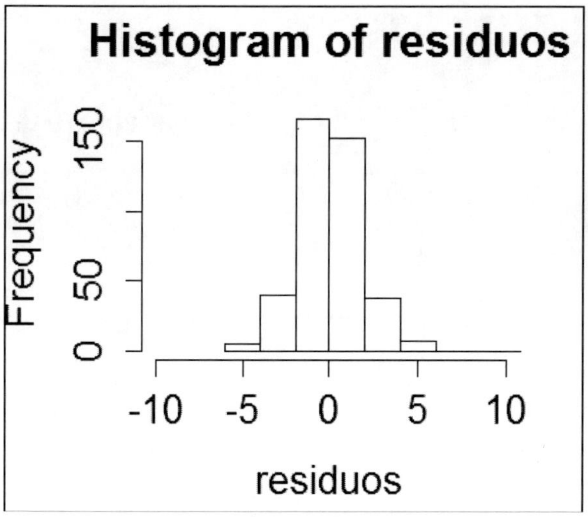

Figure 7-5

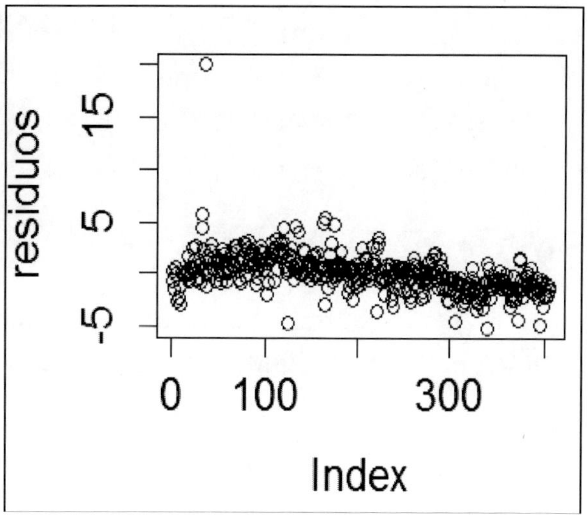

Figure 7-6

7.11 POLYNOMIAL REGRESSION

In R you can perform the polynomial regression using the following syntax:

lm(y~x+I(x^2)+I(x^3)+...)

As an example we make the regression of the consumption against the first three powers of the power.

```
> rp=lm(consumption~cv+I(cv^2)+I(cv^3))
> summary(rp)

Call:
lm(formula = consumption ~ cv + I(cv^2) + I(cv^3))

Residuals:
     Min       1Q Median      3Q     Max
-7.2530 -1.2606 -0.1914 1.1002 15.7371

Coefficients:
               Estimate Std. Error t value Pr(>|t|)
(Intercept)   4.880e+00 2.094e+00   2.330 0.02027 *
cv           -1.651e-02 5.422e-02  -0.304 0.76093
I(cv^2)       1.126e-03 4.368e-04   2.579 0.01027 *
I(cv^3)      -3.510e-06 1.099e-06  -3.195 0.00151 **
---
Meaning codes: 0 '***' 0.001 '**' 0.01 '*' 0.05 '.' 0.1 ' ' 1

Residual standard error: 2.073 on 402 degrees of freedom
Multiple R-squared: 0.7263,   Adjusted R-squared: 0.7242
F-statistic: 355.6 on 3 and 402 DF, p-value: < 2.2e-16
```

7.12 REGRESSION WITH INTERACTIONS

In R you can perform the regression containing interactions of the variables using the following syntax:

lm(y~x+z+x:z+...)

As an example we make a regression of the consumption against the power, the engine and the interaction of power and engine.

```
> rin=lm(consumption~cv+engine+cv:engine)
> summary(rin)

Call:
lm(formula = consumption ~ hp + engine + hp:engine)

Residuals:
     Min          1Q Median         3Q        Max
-5.9681 -1.1109 -0.0549 0.9243 18.2565

Coefficients:
               Estimate Std. Error t value Pr(>|t|)
(Intercept)    1,439e-01 7,401e-01   0,194      0,846
cv             8,097e-02 9,466e-03   8,553  2,54e-16 ***
engine         1.732e-03 1.999e-04   8.663  < 2e-16 ***
cv:engine     -7.239e-06 1.544e-06  -4.689  3.77e-06 ***
---
Meaning codes: 0 '***' 0.001 '**' 0.01 '*' 0.05 '.' 0.1 ' ' 1

Residual standard error: 1.954 on 402 degrees of freedom
Multiple R-squared: 0.7568,    Adjusted R-squared: 0.7549
F-statistic: 416.9 on 3 and 402 DF, p-value: < 2.2e-16
```

7.13 REGRESSION WITHOUT CONSTANT

The regression without constant is made placing as first independent variable the variable zero. The syntax will be the following:

r=lm(y~0+x1+x2+x3+....)

As an example we are going to adjust the regression model of the previous example without a constant, since the constant was not significant because it had a very high value. The syntax will be the following:

```
> rin=lm(consumption~0+cv+engine+cv:engine)
> summary(rin)
```

```
Call:
lm(formula = consumption ~ 0 + hp + engine + hp:engine)

Residuals:
    Min      1Q Median      3Q     Max
-6.0099 -1.1064 -0.0644 0.9205 18.2441

Coefficients:
            Estimate Std. Error t value Pr(>|t|)
cv         8.265e-02 3.833e-03   21.56 <2e-16    ***
engine     1.755e-03 1.619e-04   10.84 <2e-16    ***
cv:engine -7.512e-06 6.467e-07  -11.62 <2e-16    ***
---
Meaning codes: 0 '***' 0.001 '**' 0.01 '*' 0.05 '.' 0.1 ' ' 1

Residual standard error: 1.952 on 403 degrees of freedom
Multiple R-squared: 0.9735,    Adjusted R-squared: 0.9733
F-statistic: 4932 on 3 and 403 DF, p-value: < 2.2e-16
```

We observe that the individual significance of all the estimated parameters is very good (very low p-values of the T) and the joint significance is also very good (very small p-value of the F).

7.14 ADVANCED DIAGNOSTICS IN THE MULTIPLE LINEAR REGRESSION MODEL IN R

Autocorrelation	*It is detected through the Durbin Watson statistic, which must be close to the value 2. It is also detected by plotting the estimated residual against the row number. The scatter plot has to be random. There are also specific contrasts such as the Breush-Godfrey test* *It is corrected by introducing appropriate dummies in the model or by adding delays in the residues as additional independent variables. There are also other methods such as Cocharn-Orcut, Durbin and Prais-Winsten. The least-squares method can also be used for estimation.*
Heteroscedasticity	*It is detected by plotting the predicted values against the residues. If the graph is not random there is heteroscedasticity. In this case the independent variable guilty of the problem will be detected by*

	plotting all the independent variables against the estimated residues. If any graph is not random, the vriable of that graph will be the culprit of heteroscedasticity. There are also specific contrasts of heteroscedasticity as the tests of White, Ramsey, Goldfeld-Quandt, Breush-Pagan and Glesjer. *It is corrected by eliminating the guilty variable or variables from the model or by introducing appropriate dummies. It is also corrected by appropriate estimates such as the White method or the weighted least-squares method*
Multicollinearity	*It is detected by the variance inlaction factors that must be less than 10. It is also detected by the condition indices that must all be less than 30 or by the matrix of ciorrelations of the independent variables that must not have very significant values. It is also detected by the presence of a very high R2 value and little significance in the variables or by high joint significance and very low R2.* *It is corrected by reducing the independent wands to their main components and by estimating the model on the components. Once the model is estimated, the components are substituted according to the initial variables and the equation estimated in the initial variables is obtained. The chain regression method can also be used.*
Normality	*It is detected by analyzing the normality of the residues through its QQ graph, its frequency histogram or a formal contrast of normality for the residues (Kolmogrov-Smirnov, Anderson Darlin, Shapiro-Wilks, etc.)* *It is corrected by relativizing the variables in the model, using the variables in logarithms or using any trnasformation of the Box-Cox family*

The *car* library of R incorporates commands for the detection and treatment of diagnostic problems in the regression model. The most important commands have the following syntax

qqPlot	*Make the QQ chart of the residues to check the residual normality*
durbinWatsonTest	*Obtains the Durbin Watson statistic to detect the presence of autocorrelation*
outlierTest	*Bonferroni test to detect outliers*
avPlots	*Add variable graphs to detect problems*
influencePlot	*Detects influential points in the regression*
vif	*It gets the inflation factors from the variance*
crPlots	*Graphs of components plus residues to detect non-linearity*
scatterplot	*Advanced Scatter Plot*
scaterplotMatrix	*Advanced Scatterplot Matrix*
ncvTest	*Contrast of heteroscedasticity If the p-value is small there is heteroscedasticity*
spreadLevelPlot	*Creates the graph of the absolute value of the standardized residues against the adjusted values. If it is random there is no heteroscedasticity*

Next we obtain the graph of the residues for the regression without constant (Figure 7-7).

```
> qqPlot(rin)
```

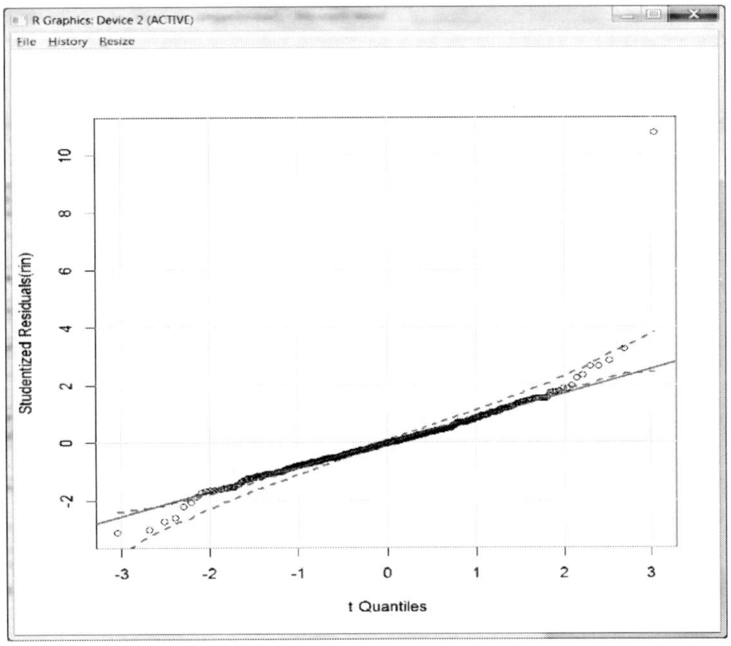

figure 7-7

The following syntax is used to obtain the Durbin Watson statistic:

```
> durbinWatsonTest(rin)
 lag Autocorrelation D-W Statistic p-value
 1         0.3758075        1.245947          0
 Alternative hypothesis: rho ! = 0
```

It is observed that the value 1.24 of the D-W statistic allows suspecting the presence of autocorrelation, certified by the p-value zero of the formal contrast of Durbin Watson.

To analyze the heteroscedasticity of the model we use the following syntax:

```
> ncvTest(rin)
Non-constant Variance Score Test
Variance formula: ~ fitted.values
Chisquare = 2.298479     Df = 1        p = 0.1295008
```

The high p-value of the contrast indicates that there is no heteroscedasticity in the model.

The modulus of the standardized residues against the adjusted values is random (Figure 7-8), which confirms the absence of heteroscedasticity problems.

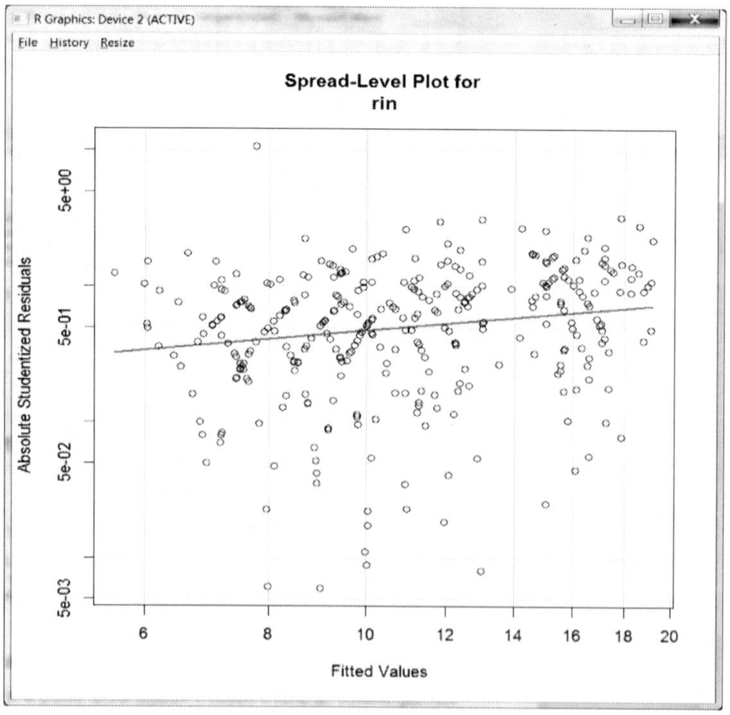

Figure 7-8

7.15 GENERALIZED LINEAR MODELS IN R. LOGIT AND PROBIT MODELS

In R you can adjust advanced models through the glm command whose general syntax is as follows:

glm(formula, family=family(link=function))

The families and their default links that can be used are the following:

binomial	*link="logit"*
gaussian	*link="identity"*
gamma	*link="inverse"*
inverse.gaussian	*link="1/mu^2"*
poisson	*link="log"*
quasi	*link="identity", variance="constant"*
quasibinomial	*link="logit"*
qasipoisson	*link="log"*

It is possible to obtain the following information regarding a GLM model:

summary(r)	*Displays detailed results of the fitted model*
coefficients(r)	*Lists the adjusted parameters of the model*
confint(r)	*Lists the 95% confidence intervals for the estimated model parameters*
plot(hatvalues(model))	*Lists the predicted values in the adjusted model*
residuals(r)	*List the residues of the adjusted model*
anova(r)	*List the ANOVA table of the adjusted model*
plot(rstudent(model))	*Graphs the residual studied*
influencePlot(model)	*Detects influential points*
plot(cooks.distance(model))	*Detects atypical residues*
plot(r)	*Displays graphical diagnostics for the fitted model*
predict(r)	*Predicts values with the model fitted*

As a first example ***we set a logit model*** that predicts the probability of car breakdowns as a function of consumption, power and engine.

```
> logistics=glm(derivative~consumption+cv+engine, family=bino
mial(link="logit"))
> summary(logistics)

Call:
glm(formula = derivative ~ consumption + hp + engine, family
= binomial(link = "logit"))

Deviance Residuals:
      Min          1Q      Median          3Q         Max
-3.06558 -0.00061    0.00316    0.02105    1.74268

Coefficients:
               Estimate Std. Error z value Pr(>|z|)
(Intercept)      25.5108094 4.8037094    5.311 1.09e-07 ***
consumption      -0.5701033 0.2210108 -2.580 0.009894 **
cv               -0.0628101 0.0223905 -2.805 0.005028 **
engine           -0.0023797 0.0006716 -3.543 0.000395 ***
---
Meaning codes: 0 '***' 0.001 '**' 0.01 '*' 0.05 '.' 0.1 ` ' 1

(Dispersion parameter for binomial family taken to be 1)

     Null deviance: 470.351 on 405 degrees of freedom
Residual deviance: 50.787 on 402 degrees of freedom
AIC: 58,787

Number of Fisher Scoring iterations: 9
```

We observe that the adjustment is quite good.

To ***fit a probit model to*** the previous case we use the following syntax:

```
> probability=glm(derivative~consumption+cv+motor, family=bin
omial(link="probit"))
```

The output is as follows:

```
Coefficients:
(Intercept)          engine hp        consumption
 13.702971      -0.319538      -0.034384      -0.001226

Degrees of Freedom: 405 Total (i.e. Null); 402 Residual
Null Deviance:      470.4
Residual Deviance: 51.7        AIC: 59.7
```

The complete output is shown below:

```
> summary(probability)

Call:
glm(formula = derivative ~ consumption + hp + engine, family
= binomial(link = "probit"))

Deviance Residuals:
 Min          1Q      Median          3Q          Max
-2.91567    0.00000     0.00001     0.00297     1.73371

Coefficients:
               Estimate Std. Error z value Pr(>|z|)
(Intercept)  13,7029706 2,3908062    5,732 9,95e-09 ***
consumption  -0.3195378 0.1034361 -3.089 0.00201 **
cv           -0.0343841 0.0112141 -3.066 0.00217 **
engine       -0.0012257 0.0002993 -4.096 4.21e-05 ***
---
Meaning codes: 0 '***' 0.001 '**' 0.01 '*' 0.05 '.' 0.1 ' ' 1

(Dispersion parameter for binomial family taken to be 1)

     Null deviance: 470.351 on 405 degrees of freedom
Residual deviance: 51.705 on 402 degrees of freedom
AIC: 59,705

Number of Fisher Scoring iterations: 25
```

We observe that the adjustment is quite good.

For the ***multinomial logistic regression,*** the *multinom* command of the *mlogit* package (or the *nnet* package) is used. First, the data frame containing the data of the variables involved in the model is generated.

```
> data=data.frame(origin,consumption,cv,engine)
```

Secondly, the regression is performed:

```
> logitmulti=multinom(origin~consumption+cv+engine, data=data)
# weights: 15 (8 variable)
initial value 446.036589
iter 10 value 222.491373
final value 221.602346
converged
```

The results of the adjustment are shown below:

```
> summary(logitmulti)
Call:
multinom(formula = origin ~ consumption + hp + engine, data =
data)

Coefficients:
        (Intercept)         engine hp        consumption
Europe     2.595916 -0.04788413 0.04010493 -0.002878232
Japan      4.933121 -0.43091353 0.09204207 -0.004722537

Std. Errors:
        (Intercept)         engine hp      consumption
Europe 0.004900491 0.1069073 0.01476321 0.0005001542
Japan  0.007639659 0.1204694 0.01862427 0.0006471102

Residual Deviance: 443,2047
AIC: 459.2047
```

For the ***ordered logistic regression the*** *polr* command of the *MASS* package is used. The syntax is the following:

```
> logitordenado=polr(formula = cylinder ~ acel + consumo + cv
+ motor + peso,
+       data = data, method = "logistic")
Warning: glm.fit: fitted probabilities numerically 0 or 1 occ
urred
> summary(logitordinated)

Re-fitting to get Hessian

Warning in sqrt(diag(vc)) : NaNs produced
Call:
polr(formula = cylinder ~ acel + consumption + cv + engine +
weight,
 data = data, method = "logistic")

Coefficients:
 Value Std. Error t value
acel          0.163035         NaN      NaN
consumption 0.182883     0.13211    1.384
cv            0.046955     0.01317    3.566
engine        0.006049         NaN      NaN
weight       -0.006860         NaN      NaN

Intercepts:
```

```
                          Value    Std. Error t value
3 cylinders|4 cylinders 8,1144       NaN          NaN
4 cylinders|5 cylinders 18.3219 0.6786      26.9985
5 cylinders|6 cylinders 18.6929 0.6782      27.5619
6 cylinders|8 cylinders 28,7806     NaN          NaN

Residual Deviance: 162,9483
AIC: 180,9483
```

7.16 MULTIPLE REGRESSION MODELS IN R THROUGH MENUS

The *Fit Models* option from the *Statistics* menu (Figure 7-9) allows you to *fit* linear models, generalized linear models, logit and probit models, multinomial logit models, and ordered logit models.

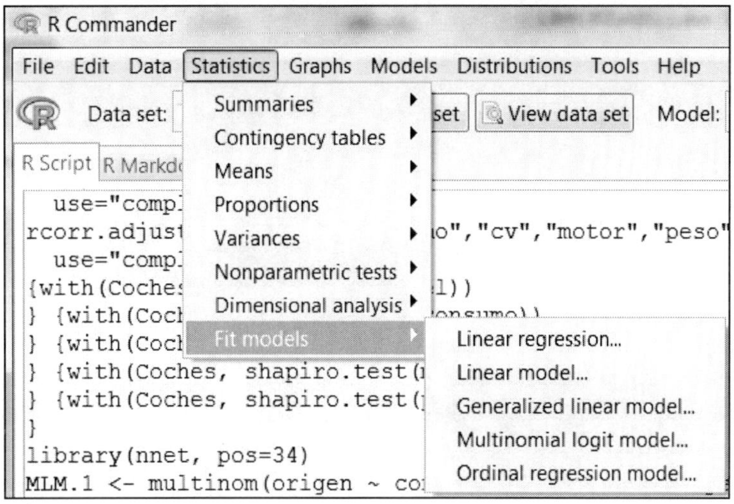

Figure 7-9

The Linear Regression option allows you to adjust the linear model of multiple regrsion. In Figure 7-10 the response variable and the explanatory variables are selected.

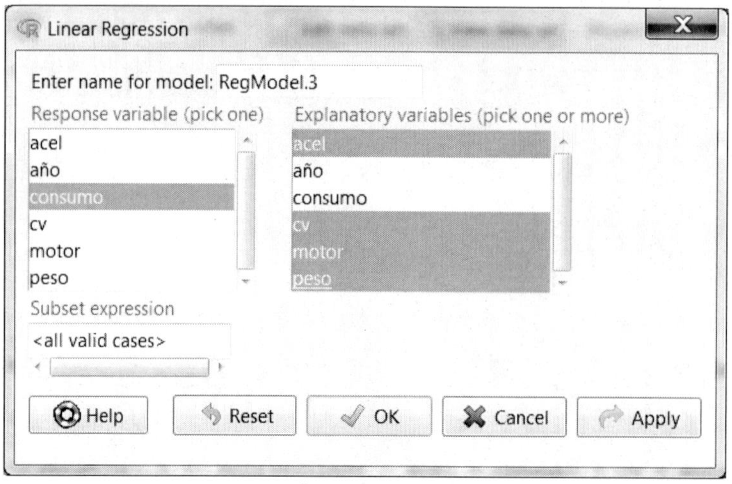

Figure 7-10

Clicking OK results in the following setting:

```
> RegModel.3 <- lm(consumption~acel+cv+engine+weight, data=Ca
rs)

> summary(RegModel.3)

Call:
lm(formula = consumption ~ acel + cv + engine + weight, data
= cars)

Residuals:
     Min       1Q Median       3Q      Max
-5.1265 -1.1090 -0.0445 0.8910 20.0902

Coefficients:
              Estimate Std. Error t value Pr(>|t|)
(Intercept)  0.3331632 1.1246075   0.296   0.7672
acel         0.0314109 0.0569507   0.552   0.5816
cv           0.0440587 0.0076345   5.771 1.58e-08 ***
engine       0.0003215 0.0001846   1.741   0.0824 .
weight       0.0048810 0.0010929   4.466 1.04e-05 ***
---
Meaning codes: 0 '***' 0.001 '**' 0.01 '*' 0.05 '.' 0.1 ' ' 1

Residual standard error: 1.94 on 401 degrees of freedom
Multiple R-squared: 0.761,    Adjusted R-squared: 0.7587
F-statistic: 319.3 on 4 and 401 DF, p-value: < 2.2e-16
```

The adjustment fails because of the non-significance of the constant and the *acceler* variable (very high p-values). We remove the acceler variable from the model and adjust it again (Figure 7-11)

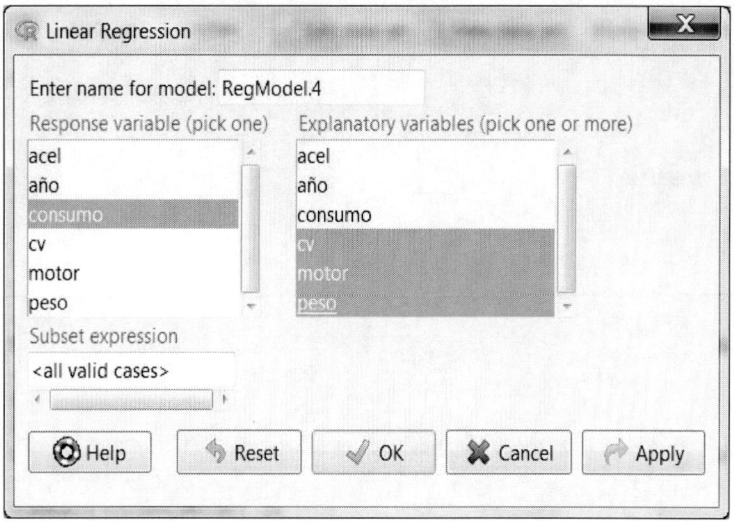

Figure 7-11

```
> RegModel.4 <- lm(consumption~cv+engine+weight, data=Cars)

> summary(RegModel.4)

Call:
lm(formula = consumption ~ hp + engine + weight, data = cars)

Residuals:
     Min       1Q Median       3Q      Max
-5.1803 -1.1196 -0.0541 0.8530 20.0023

Coefficients:
              Estimate Std. Error  t value Pr(>|t|)
(Intercept)   0.8762787 0.5427383   1.615     0.107
cv            0.0412840 0.0057376   7.195 3.07e-12 ***
engine        0.0003043 0.0001819   1.674     0.095 .
weight        0.0051719 0.0009563   5.408 1.10e-07 ***
---
Meaning codes: 0 '***' 0.001 '**' 0.01 '*' 0.05 '.' 0.1 ' ' 1

Residual standard error: 1.938 on 402 degrees of freedom
Multiple R-squared: 0.7609,   Adjusted R-squared: 0.7591
F-statistic: 426.3 on 3 and 402 DF, p-value: < 2.2e-16
```

Now the result already indicates a good fit.

Once the model is fixed and estimated, it will be possible to perform diagnostic tasks through the options in the *Models* menu (Figure 7-12).

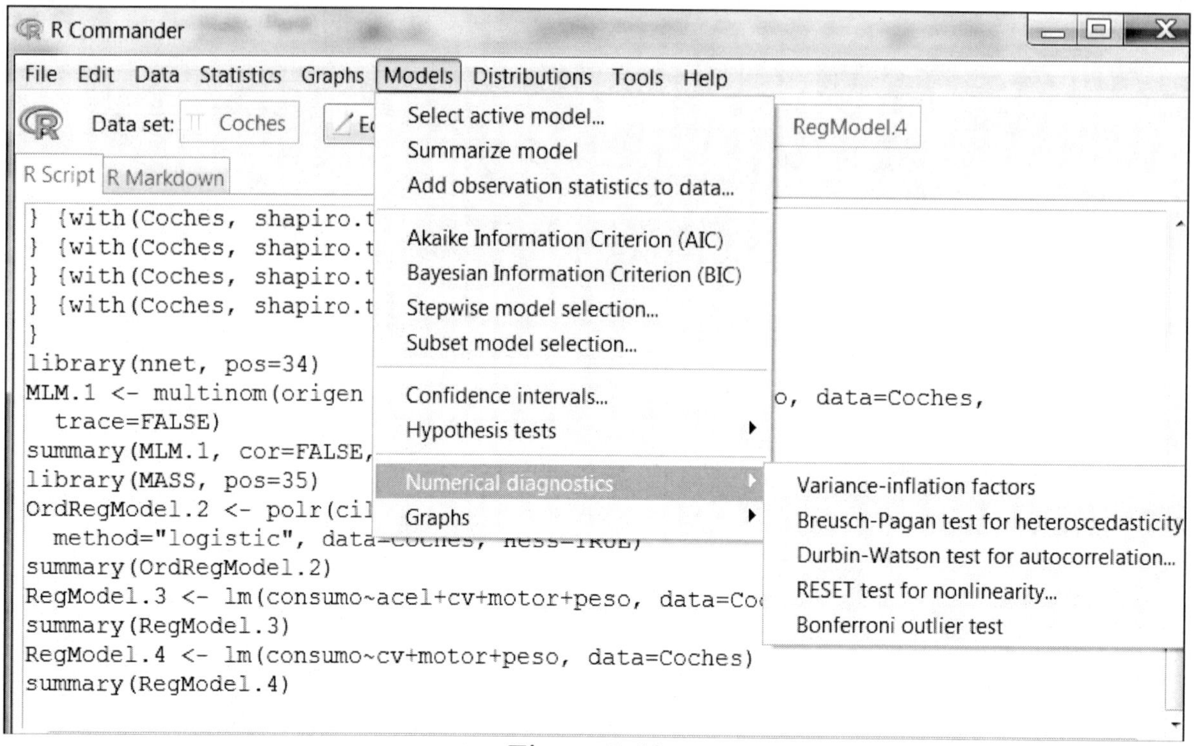

Figure 7-12

After selecting as active model the *RegModel.4 model* with the option *Select active model,* we can ask for a summary of it with the option *Summarize model.* The output is the following:

```
> summary(RegModel.4)

Call:
lm(formula = consumption ~ hp + engine + weight, data = cars)

Residuals:
     Min       1Q Median      3Q     Max
-5.1803 -1.1196 -0.0541 0.8530 20.0023

Coefficients:
              Estimate Std. Error t value Pr(>|t|)
(Intercept) 0.8762787 0.5427383    1.615    0.107
cv          0.0412840 0.0057376    7.195 3.07e-12 ***
engine      0.0003043 0.0001819    1.674    0.095 .
weight      0.0051719 0.0009563    5.408 1.10e-07 ***
---
Meaning codes: 0 '***' 0.001 '**' 0.01 '*' 0.05 '.' 0.1 ' ' 1

Residual standard error: 1.938 on 402 degrees of freedom
Multiple R-squared: 0.7609,   Adjusted R-squared: 0.7591
F-statistic: 426.3 on 3 and 402 DF, p-value: < 2.2e-16
```

A good fit is observed.

The options *Akaike Infirmation Criterium, Bayesian Information Criterium* and *Confidence* intervals allow to obtain the AKAIKE and BIC information statistics for the predictive capacity of the model and the confinement intervals for the estimated parameters of the model at the desired confidence level. The output is as follows:

```
> AIC(RegModel.4)
[1] 1695.379

> BIC(RegModel.4)
[1] 1715.411

> Confint(RegModel.4, level=0.95)
                  Estimate          2.5 %         97.5
(Intercept)   0.8762786722 -1.906812e-01 1.9432385272
cv            0.0412840065  3.000464e-02 0.0525633722
engine        0.0003043373 -5.316155e-05 0.0006618361
weight        0.0051718878  3.291844e-03 0.0070519317
```

The options *Variance inflaction factors, Breush Pagan test for Heteroscedasticity, Durbin Watson test for Autocorrelation, RESET test for nonlinearity* and *Bonferroni outliers test* contrast multicollinearity, heteroscedasticity, autocorrelation, nonlinearity and presence of outliers in the model respectively. The output is as follows:

```
> vif(RegModel.4)
       cv      engine       weight
 5.238454   10.600106     7.914785
```

The variance inflation factors are all less than 10, indicating that there are no problems of multicollinearity.

```
> library(zoo, pos=36)

> library(lmtest, pos=36)

> bptest(consumption ~ hp + motor + weight, varformula = ~
+    fitted.values(RegModel.4), studentize=FALSE, data=Cars)

        Breusch-Pagan test

data: consumption ~ hp + engine + weight
BP = 4.0059, df = 1, p-value = 0.04534
```

The Breush Pagan contrast has a p-value at the border of 0.05, indicating that there may or may not be heteroscedasticity.

```
> dwtest(consumption ~ hp + engine + weight, alternative="two
.sided", data=Cars)

        Durbin-Watson test

data: consumption ~ hp + engine + weight
DW = 1.227, p-value = 3.01e-15
alternative hypothesis: true autocorrelation is not 0
```

The Durbin Watson statistic has a low value (1.2 is far from 2) and a very sm all p-value, which indicates there is residual self-correlation.

```
> resettest(consumption ~ hp + engine + weight, power=2:3, ty
pe="regressor",
+     data=Cars)

        RESET test

data: consumption ~ hp + engine + weight
RESET = 22,523, df1 = 6, df2 = 396, p-value < 2.2e-16
```

Ramsey's RESET contrast has a very low p-value, indicating that there are no linearity problems.

```
> outlierTest(RegModel.4)
    rstudent unadjusted p-value Bonferonni p
35 12.51266         1.4829e-30   6.0206e-28
```

The Bonferronni contrast has a very low p-value, indicating that there are infl uential outliers in the regression.

The *Graphs* option on the *Model* menu allows you to perform a graphical diagnosis of the model (Figure 7-13).

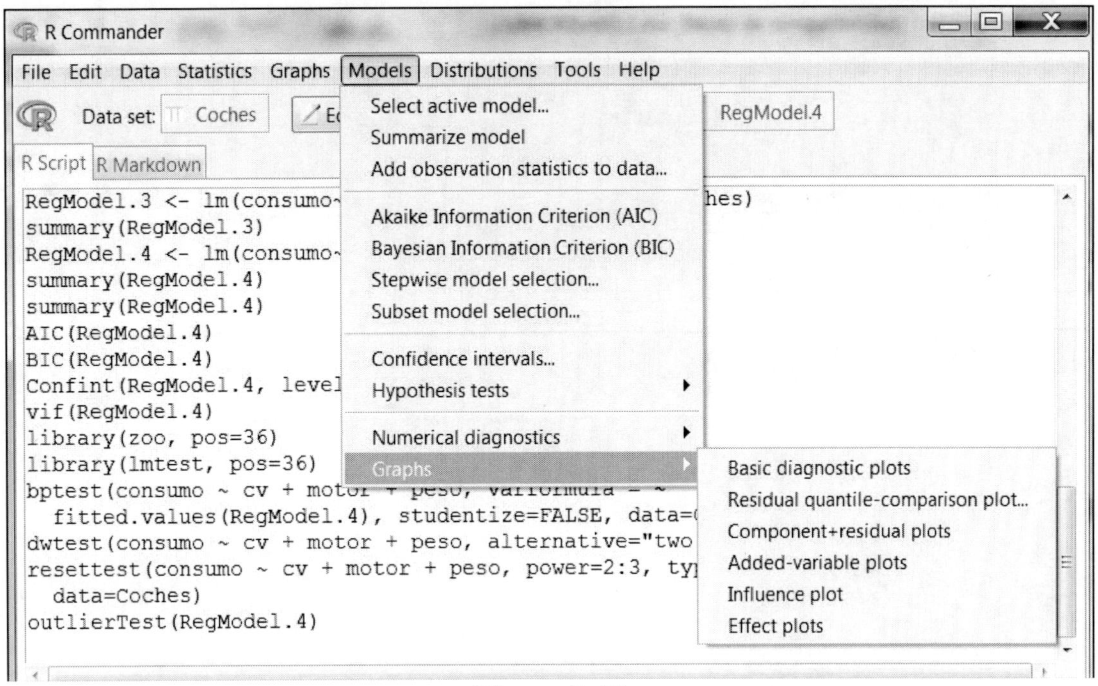

Figure 7-13

The *Basic Diagnostic Plots* option performs the graphic analysis of the residues (Figure 7-14).

The *Residual Quantile Comparison Plot* option performs a QQ chart for residual normality (Figure 7-15). It should be set to the diagonal of the first quadrant.

The *Component+residual plots* option performs a graph of components plus residues against the independent variables to contrast the linearity (Figure 7-16). The point clouds must fit on lines parallel to the diagonal.

The *Added-variable plots* option performs a graph of predicted values against the independent variables to contrast the linearity (Figure 7-17). The point clouds must be set to lines parallel to the diagonal.

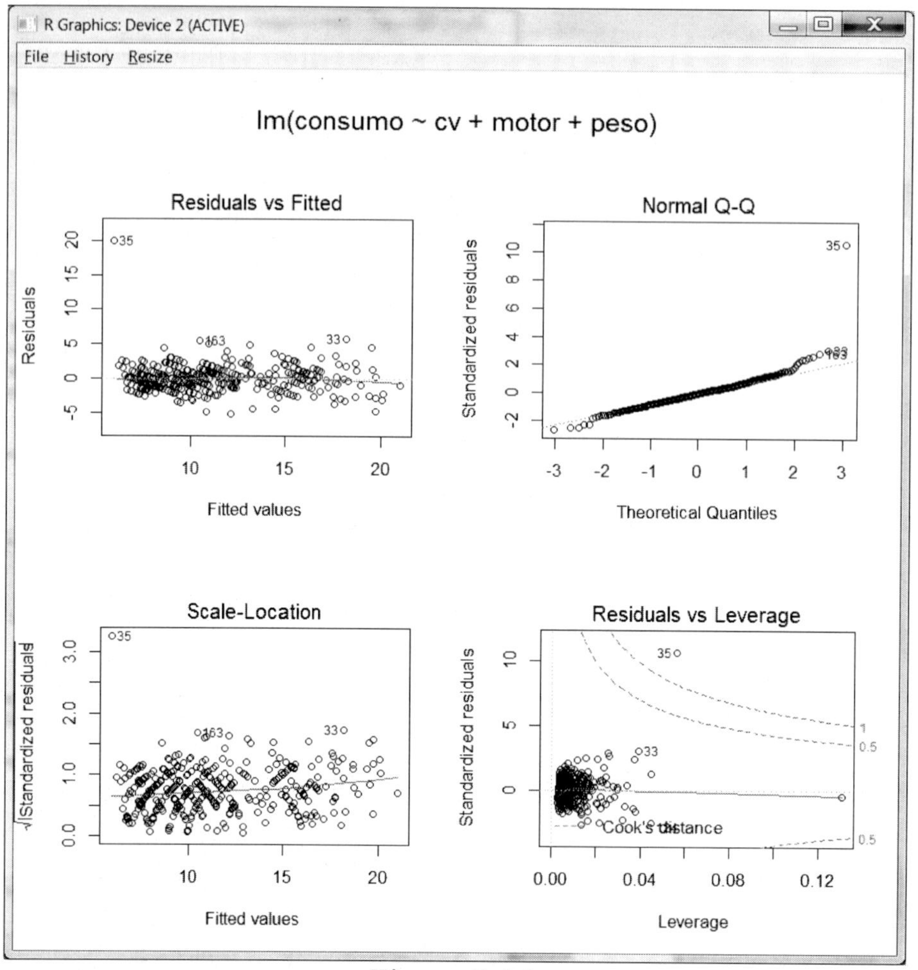

Figure 7-14

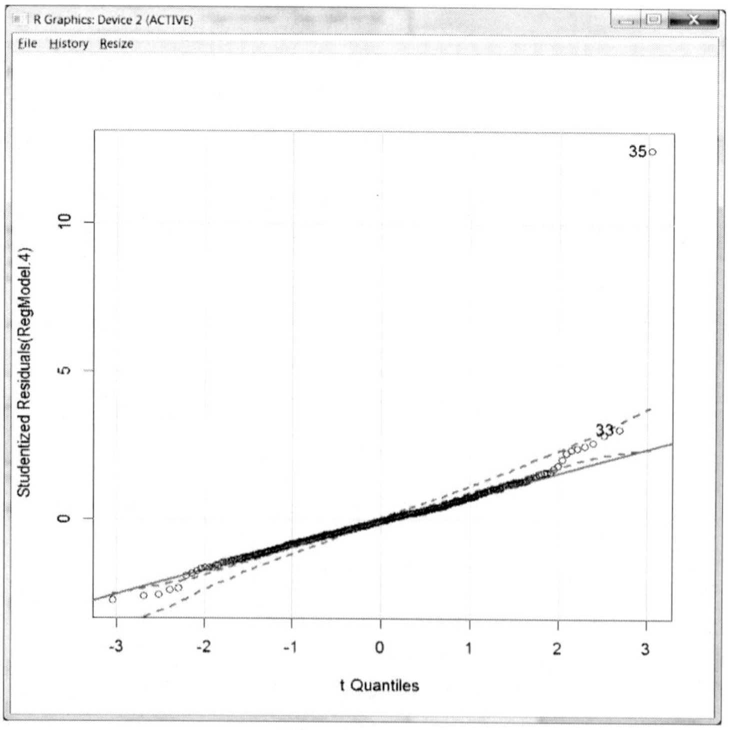

Figure 7-15

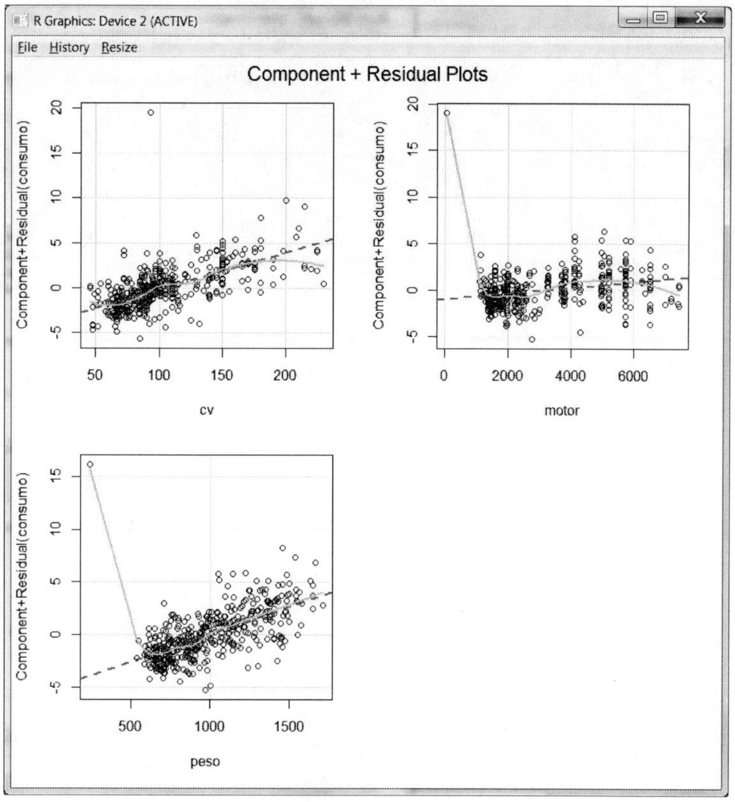

Figure 7-16

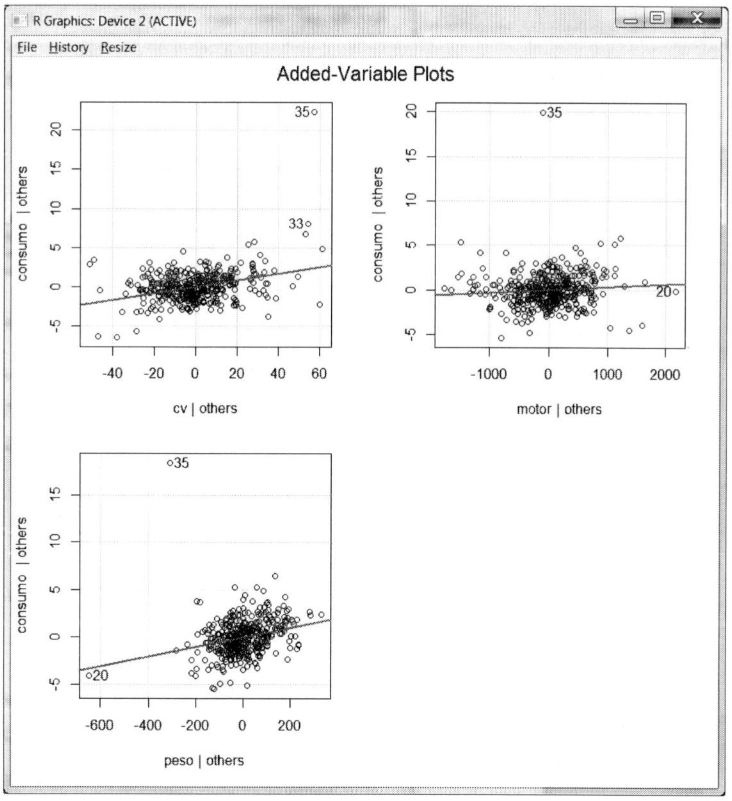

Figure 7-17

The *Influence plots* option performs a graph to detect influent or outliers (Figure 7-17a). It can be seen that the value in row 35 of the database causes an outlier.

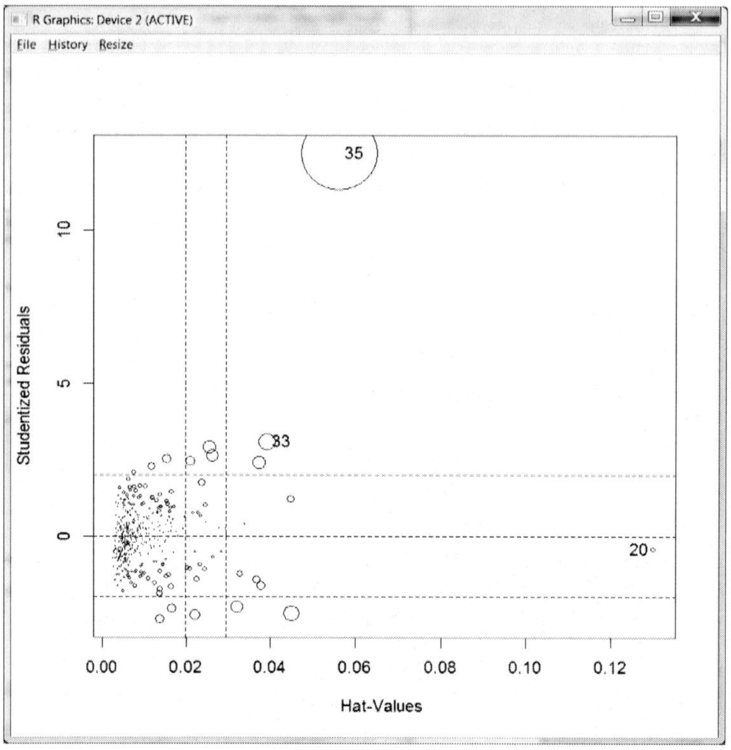

Figure 7-17a

The *Effect plots* option *plots* each independent variable against the dependent variable (Figure 7-18).

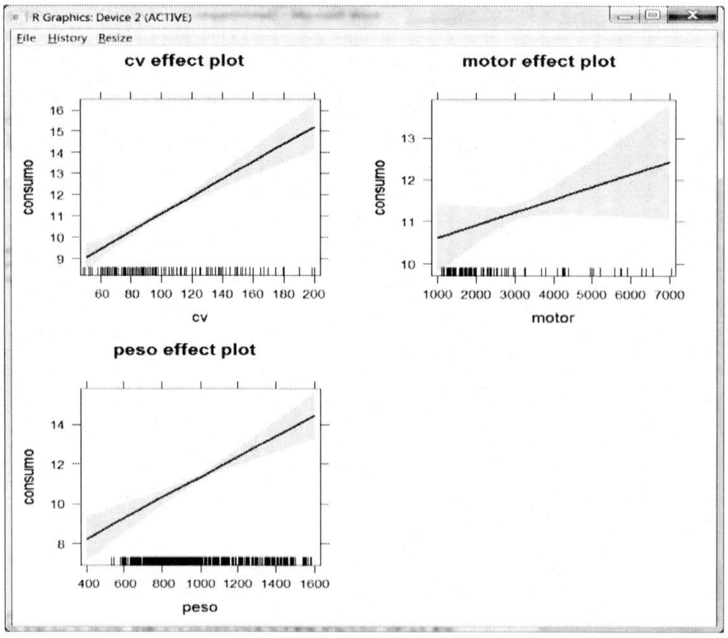

Figure 7-18

The *Stepwise model selection* option from the *Model* menu (Figure 7-13) brings up the screen in Figure 7-19, which allows you to perform a stepwise regression forward, backward, or combined.

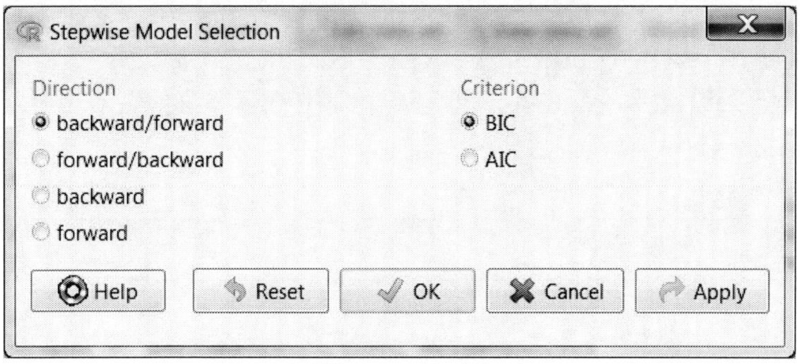

Figure 7-19

The output is as follows:

```
> stepwise(RegModel.4, direction='backward/forward', criterio
n='BIC')

Direction: backward/forward
Criterion: BIC

Start: AIC=557.23
consumption ~ hp + engine + weight

         Df  Sum of Sq     RSS    AIC
- engine 1     10.518   1520.2  554.04
<none>                  1509.7  557.23
- weight 1    109,834   1619.5  579.73
- cv     1    194,433   1704,1  600,41

Step: AIC=554.04
consumption ~ hp + weight

         Df Sum of Sq     RSS    AIC
<none>                 1520.2  554.04
+ engine 1      10.52  1509.7  557.23
- cv     1     333.45  1853.7  628.55
- weight 1     338.65  1858.8  629.69

Call:
lm(formula = consumption ~ hp + weight, data = cars)

Coefficients:
(Intercept)            cv          weight
    0.187677      0.046255      0.006321
```

7.17 GENERALIZED LINEAR MODELS IN R. LOGIT AND PROBIT MODELS THROUGH MENUS

The *General linear model* suboption of the *Fit models* option in the *Statistics* menu allows (Figure 7-20) to adjust generalized linear models, *logit, probit, clog* and others. Filling in the input screen as shown in Figure 7-21 you get the output for the ***binomial logit model*** shown on the next page.

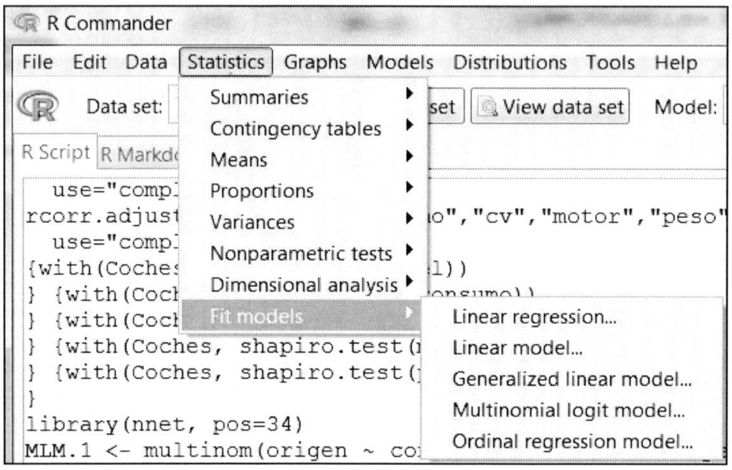

Figure 7-20

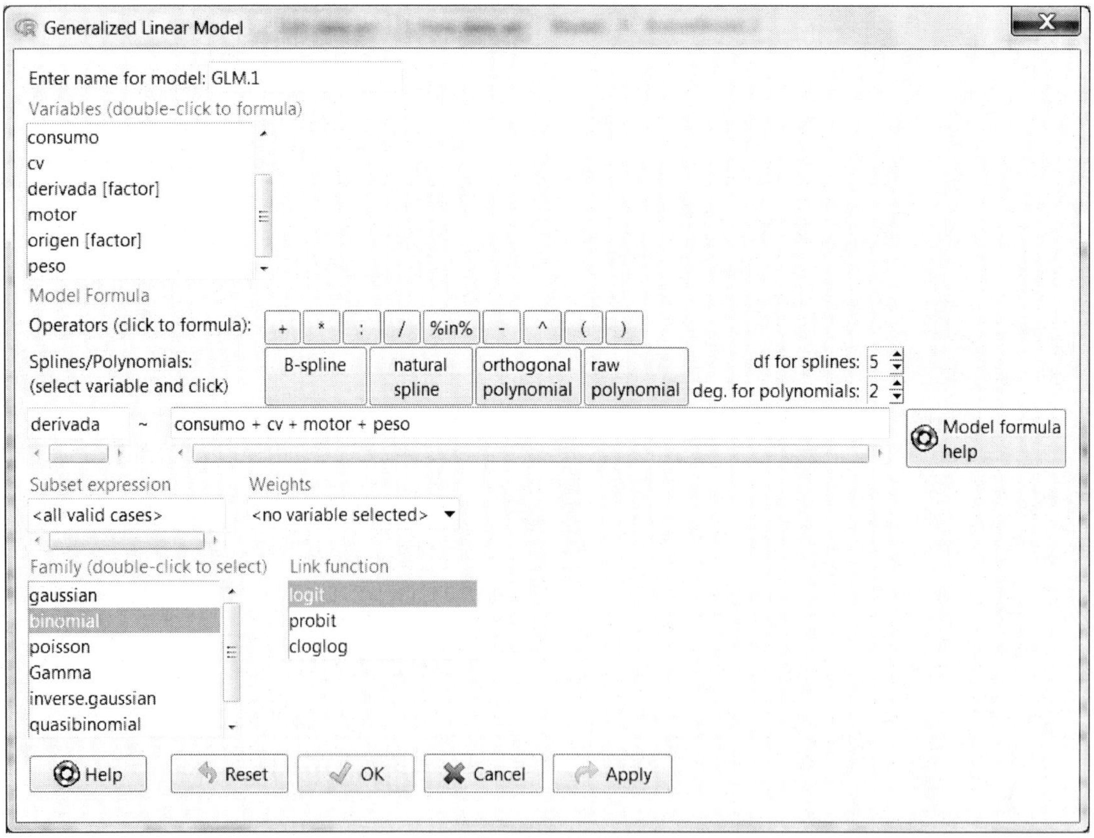

Figure 7-21

```
Rcmdr> GLM.1 <- glm(derivative ~ consumption + cv + engine +
weight, family=binomial(logit),
Rcmdr+       data=Data)

Rcmdr> summary(GLM.1)

Call:
glm(formula = derivative ~ consumption + hp + engine + weight
, family = binomial(logit),
     data = Data)

Deviance Residuals:
      Min          1Q      Median          3Q         Max
-3.02943 -0.00019    0.00192    0.01133    1.71982

Coefficients:
              Estimate Std. Error z value Pr(>|z|)
(Intercept)  23.886284    5.014976   4.763 1.91e-06 ***
consumption  -0.689723    0.206173 -3.345 0.000822 ***
cv           -0.075638    0.024650 -3.069 0.002151 **
engine       -0.003919    0.001110 -3.530 0.000415 ***
weight        0.009733    0.004265    2.282 0.022490 *
---
Meaning codes: 0 '***' 0.001 '**' 0.01 '*' 0.05 '.' 0.1 ' ' 1

(Dispersion parameter for binomial family taken to be 1)

     Null deviance: 470.35 on 405 degrees of freedom
Residual deviance: 44.63 on 401 degrees of freedom
AIC: 54.63

Number of Fisher Scoring iterations: 9
```

It is observed that the adjustment of the binomial logit model that explains the probability of failure of a car according to its consumption, power, engine and weight is correct since the p-values of the estimated parameters are very low.

If we want to make the above adjustment through a ***probit model,*** we fill in the input screen as shown in Figure 7-22. Clicking on *OK* stops the output of the next page.

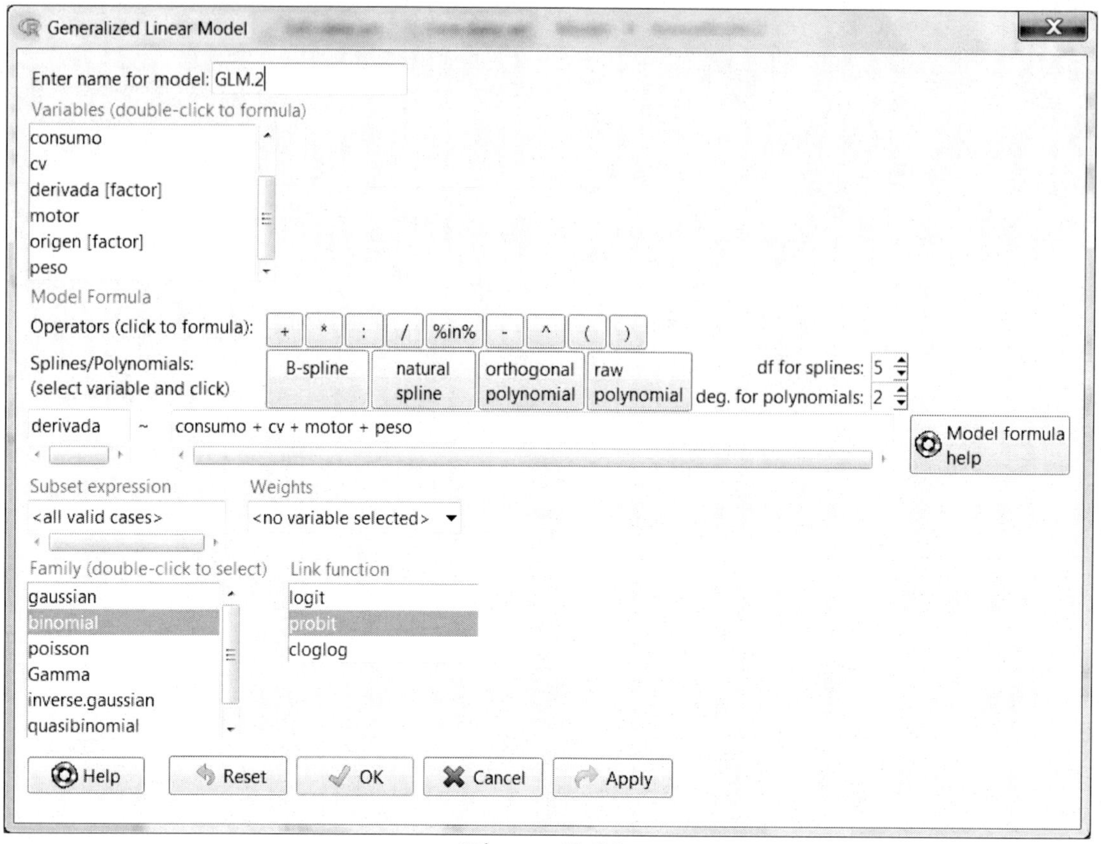

Figure 7-22

```
Rcmdr> GLM.2 <- glm(derived ~ consumption + hp + engine + wei
ght, family=binomial(probit), data=Data)
RcmdrMsg: [4] AVISO: Warning: glm.fit: fitted probabilities n
umerically 0 or 1 occurred

Rcmdr> summary(GLM.2)

Call:
glm(formula = derivative ~ consumption + hp + engine + weight
, family = binomial(probit),
     data = Data)

Deviance Residuals:
      Min          1Q      Median          3Q         Max
-3.06907    0.00000    0.00000    0.00055    1.68325

Coefficients:
               Estimate Std. Error z value Pr(>|z|)
(Intercept) 12,7396040 2,4085454    5,289 1,23e-07 ***
consumption -0.3529930 0.1058517 -3.335 0.000854 ***
cv          -0.0403722 0.0128115 -3.151 0.001626 **
engine      -0.0020675 0.0005446 -3.796 0.000147 ***
weight       0.0049110 0.0021290    2.307 0.021070
```

```
---
Meaning codes: 0 '***' 0.001 '**' 0.01 '*' 0.05 '.' 0.1 ' ' 1

(Dispersion parameter for binomial family taken to be 1)

    Null deviance: 470.351 on 405 degrees of freedom
Residual deviance: 45.021 on 401 degrees of freedom
AIC: 55,021

Number of Fisher Scoring iterations: 11
```

It is observed that the adjustment of the binomial logit model that explains the probability of failure of a car according to its consumption, power, engine and weight is correct since the p-values of the estimated parameters are very low.

To estimate a ***multinomial logit*** *model* that explains the probability of a car's origin according to its consumption, power, engine and weight, let's fill in the input screen of the *Multinomial logit model* sub-option of the *Fit Models* option in the *Statistics* menu (Figure 7-20) as shown in Figure 7-23. Clicking on OK you get the output shown on the next page.

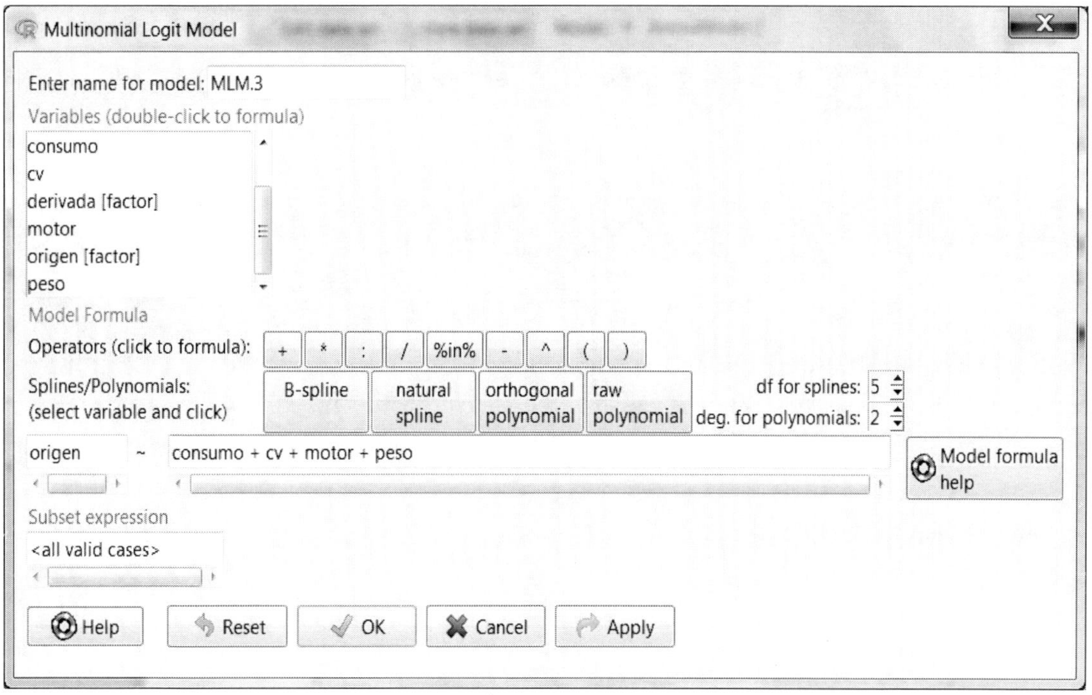

Figure 7-23

```
Rcmdr> MLM.3 <- multinom(origin ~ consumption + hp + engine +
weight, data=Data, trace=FALSE)

Rcmdr> summary(MLM.3, cor=FALSE, Wald=TRUE)
```

```
Call:
multinom(formula = origin ~ consumption + hp + engine + weigh
t, data = data,
     trace = FALSE)

Coefficients:
        (Intercept)     consumption            cv           motor
weight
Europe -0.7723152 -0.1285933 0.01986355 -0.006161812 0.014892
476
Japan     3.5844798 -0.4812759 0.09127008 -0.005770095 0.0048
77201

Std. Errors:
        (Intercept)    consumption            cv           motor
weight
Europe 0.002474577 0.1035906 0.01761006 0.0009432347 0.002530
233
Japan 0.003522428 0.1125036 0.01971565 0.0009911528 0.0026023
07

Value/SE (Wald statistics):
        (Intercept)    consumption          cv      motor
weight
Europe     -312.0999 -1.241361 1.127966 -6.532639 5.885811
Japan      1017.6163 -4.277872 4.629322 -5.821600 1.874184

Residual Deviance: 407,426
AIC: 427,426
```

To estimate the ordered logit model that explains the probability of a car's displacement as a function of its consumption, power, engine and weight, fill in the input screen of the Ordinal regression model sub-option of the Fit models option in the Statistics menu (Figure 7-20) as shown in Figure 7-24. Pressing OK gives the output shown on the next page.

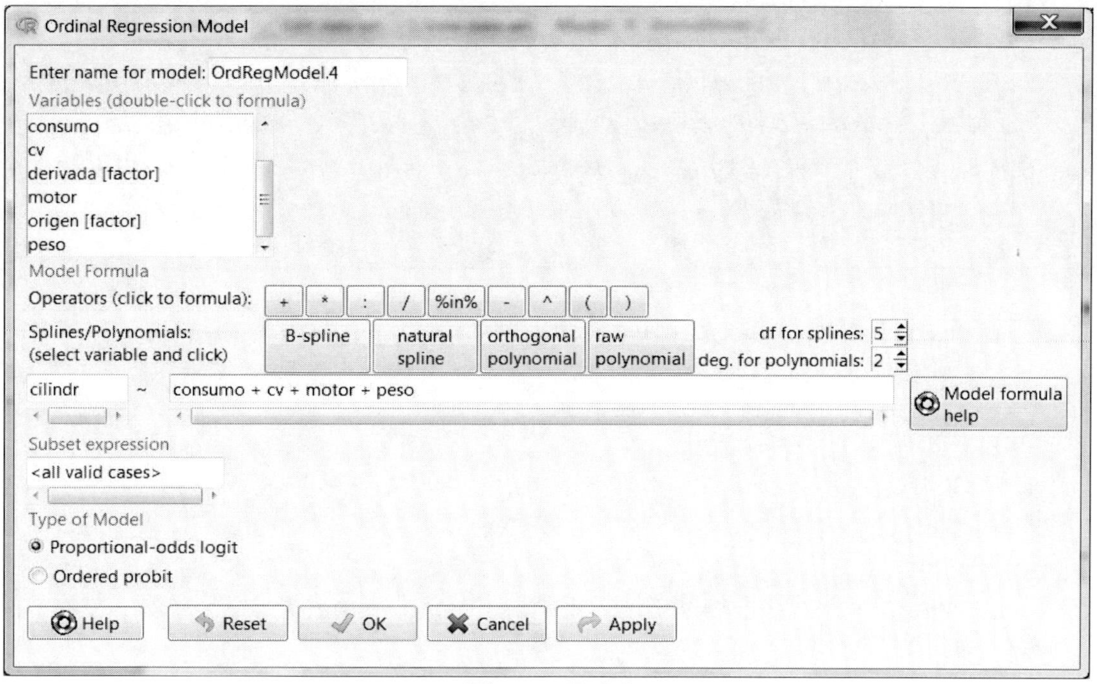

Figure 7-24

```
Rcmdr> OrdRegModel.4 <- polr(cylinder ~ consumption + hp + en
gine + weight, method="logistic",
Rcmdr+     data=Data, Hess=TRUE)
RcmdrMsg: [5] AVISO: Warning: glm.fit: fitted probabilities n
umerically 0 or 1 occurred

Rcmdr> summary(OrdRegModel.4)
Call:
polr(formula = cylinder ~ consumption + hp + engine + weight,
data = data,
     Hess = TRUE, method = "logistic")

Coefficients:
             Value Std. Error t value
consumption 0.204597    0.139318 1.4686
cv         0.027051    0.032173 0.8408
engine      0.005813    0.001424 4.0808
weight      -0.004828    0.002190 -2.2045

Intercepts:
 Value    Std. Error t value
3 cylinders|4 cylinders 5.1552 0.1643     31.3759
4 cylinders|5 cylinders 15.3329 0.7421     20.6625
5 cylinders|6 cylinders 15.7061 0.7406     21.2083
6 cylinders|8 cylinders 25,5940 3,4050      7,5166

Residual Deviance: 164,1497
AIC: 180,1497
```

Exercise 7-1. The file viviendas.sav contains information about neighborhoods of a city, price of the houses, price of the land, value of the improvements and ratio between real price of the house and appraised price. The aim is to estimate a linear regression model that explains the price of the house according to the price of the land, the value of the improvements and the ratio between the real price of the house and the appraised price

First we read the file as a list and enable its variables for analysis.

```
> library(foreign)
> viviendas=read.spss("G:/R/data/viviendas.sav", use.value.la
bels=TRUE)
> attach(dwellings)
```

Next we explore the variables of the file and its nature.

```
> summary(housing)
             Length    Class Mode
SC            2440     -none- character
neighborhood 2440     factor numeric
valterr       2440     -none- numeric
valmejor      2440     -none- numeric
valtot        2440     -none- numeric
price         2440     -none- numeric
rate          2440     -none- numeric
```

We observe that the variable SC is of type character, the variable neighborhood is a factor and the rest of the variables are numerical-

Then we perform the requested regression:

```
> reg1=lm(price~valterr+valmejor+rate)
> summary(reg1)

Call:
lm(formula = price ~ valterr + valmejor + tax)

Residuals:
     Min      1Q Median       3Q      Max
-153565 -10446    -496     8764  355496

Coefficients:
              Estimate Std. Error t value Pr(>|t|)
(Intercept)  1.022e+03 1.292e+03    0.791 0.42884
valterr      3.186e+00 5.336e-02   59.710 < 2e-16 ***
valmejor     4.682e-01 2.571e-02   18.209 < 2e-16 ***
rate         5.993e+01 2.138e+01    2.804 0.00509 **
```

```
---
Meaning codes: 0 '***' 0.001 '**' 0.01 '*' 0.05 '.' 0.1 ' ' 1

Residual standard error: 28030 on 2436 degrees of freedom
Multiple R-squared: 0.6767,    Adjusted R-squared: 0.6763
F-statistic: 1699 on 3 and 2436 DF, p-value: < 2.2e-16
```

It is observed that all the estimated parameters are significant (low p-values) except the constant (high p-value). It will be necessary to remove the constant from the model and re-estimate it. For this we use the following syntax:

```
> reg1=lm(price~0+valterr+valmejor+rate)
> summary(reg1)

Call:
lm(formula = price ~ 0 + valterr + valmejor + tax)

Residuals:
     Min       1Q Median       3Q      Max
-153724 -10127    -227     8952 356216

Coefficients:
          Estimate Std. Error t value Pr(>|t|)
valterr    3.20528    0.04738 67.648 < 2e-16 ***
valmejor   0.48037    0.02060 23.319 < 2e-16 ***
rate      58.74124   21.32115  2.755 0.00591 **.
---
Meaning codes: 0 '***' 0.001 '**' 0.01 '*' 0.05 '.' 0.1 ' ' 1

Residual standard error: 28030 on 2437 degrees of freedom
Multiple R-squared: 0.9021,    Adjusted R-squared: 0.902
F-statistic: 7484 on 3 and 2437 DF, p-value: < 2.2e-16
```

Now the estimated model already has all its significant parameters individually because all p-values of the Student T are small. In addition, the overall significance of the estimated parameters is also high because the p-value of the F of Fisher is small. The adjusted square R is worth 0.9021 which indicates that the independent variables explain perfectly to the housing price variable.

To make the graphical residual diagnosis we use the following syntax

```
> windows()
> plot(reg1)
```

Figure 7-25 shows the graph of predicted values against residues and it turns out to be random, indicating that there are no problems of heteroscedasticity. Figure

7-26 shows the residual normality QQ graph which is well adjusted to the diagonal of the first quadrant, indicating that there is residual normality. The graph of the predicted values against the square root of the standardized residuals is a random cloud (Figure 7-27), which indicates that there are no non-linearity problems. The graph of the influence against the standardized residuals does not leave any points outside the outer confinement bands (Figure 7-28), indicating that there are no influential outliers on the residuals. We have already diagnosed the model graphically.

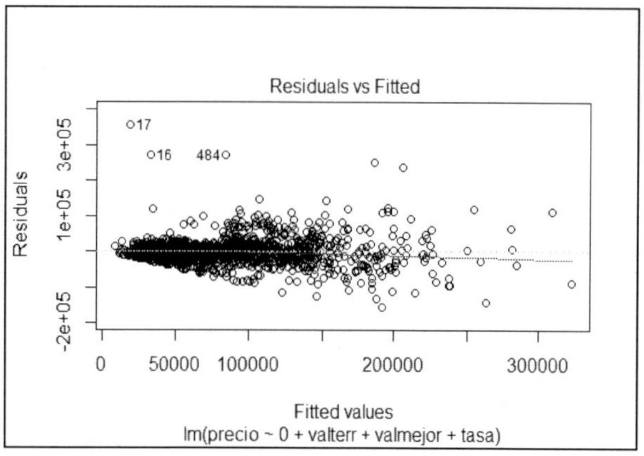

Figure 7-25

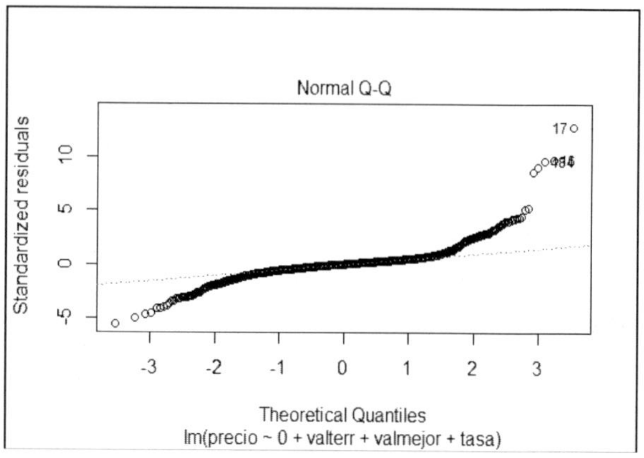

Figure 7-26

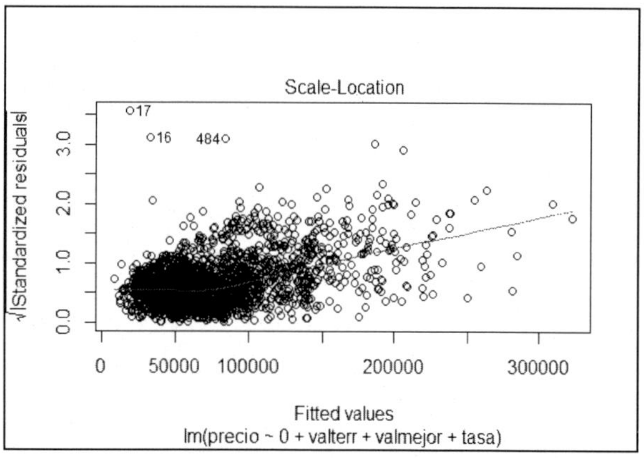

Figure 7-27

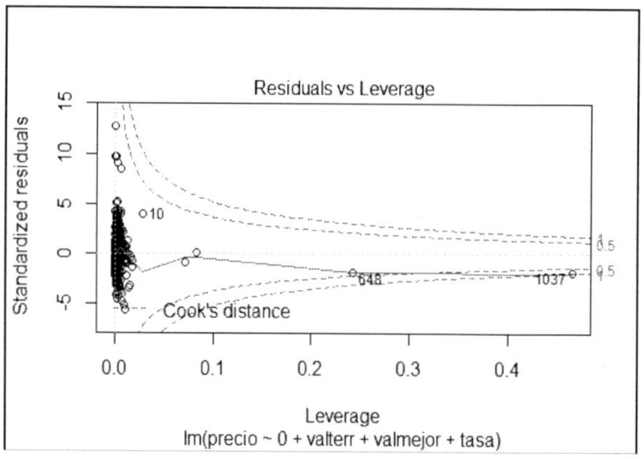

Figure 7-28

The graphical analysis of the residues can be enriched by sending the residues and the predicted values to the archive, for further analysis at a later stage.

```
> residual=residuals(reg1)
> predicted=fitted(reg1)
> hist(residual)
> plot(residual)
```

The histogram of the residuals (Figure 7-29) fits fairly well into a normal and the row number graph against the estimated residuals is random (Figure 7-30). This corroborates that there are no problems of residual normality or autocorrelation.

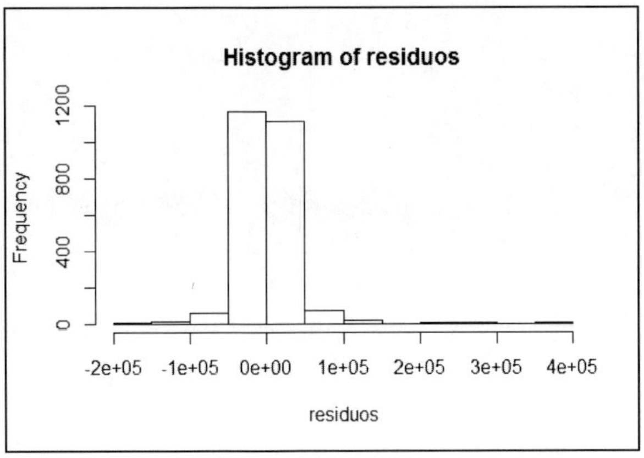

Figure 7-29

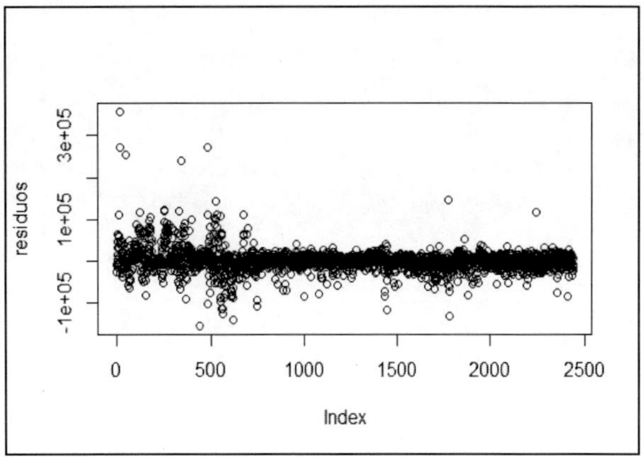

Figure 7-30

On the other hand, the graphs of the predicted values against the independent variables fit well on a line, indicating that there are no problems of linearity flats (Figures 7-30 and 7-31).

```
> plot(predicted,valterr)
> plot(predicted,valmejor)
```

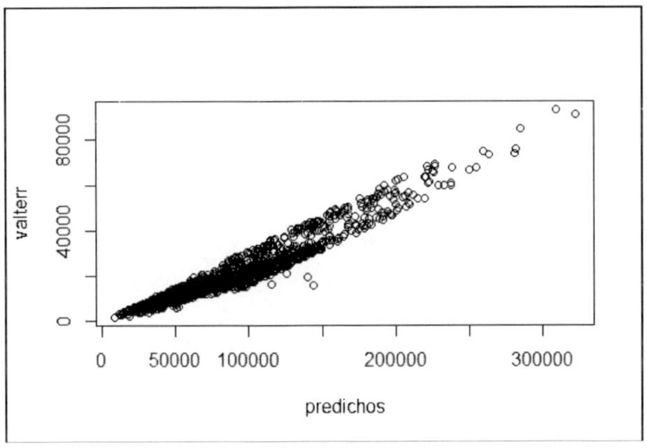

Figure 7-30

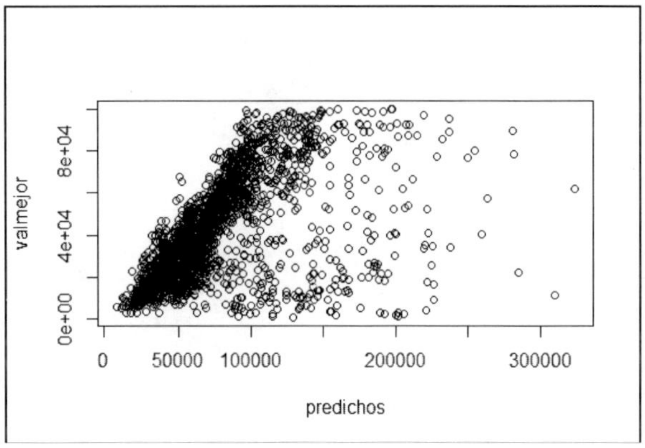

Figure 7-31

When graphing the model-independent variables against the estimated residues we observe random graphs (Figures 7-32 and 7-33), indicating that there are no problems of heterosdecasticity.

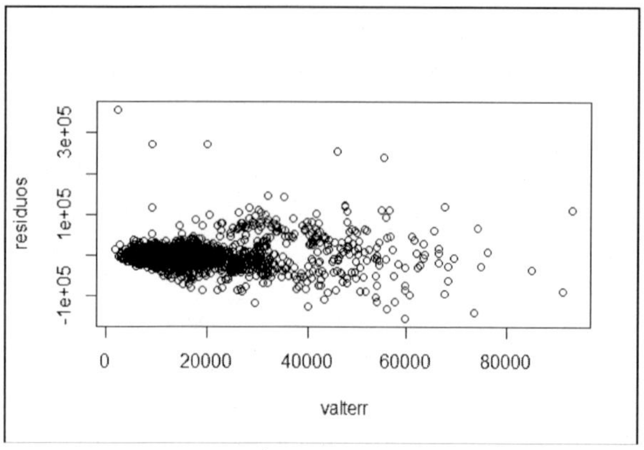

Figure 7-32

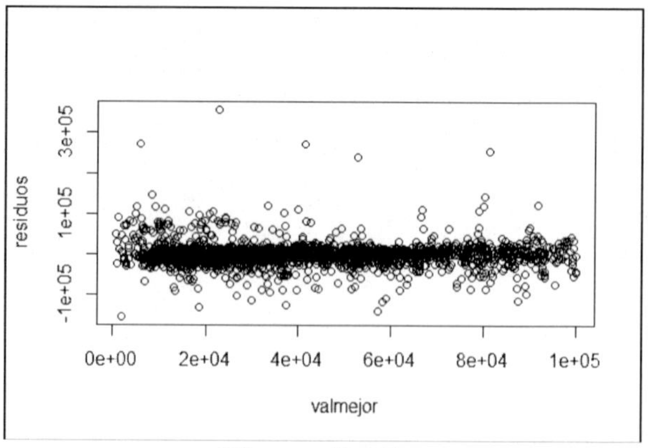

Figure 7-33

In addition, the covariance matrix of the estimated parameters has small values, indicating that the estimate may be correct.

```
> vcov(reg1)
                valterr       valmejor          rate
valterr     0.002245016 -0.0007837420   -0.006848487
valmejor   -0.000783742  0.0004243702   -0.050501563
rate       -0.006848487 -0.0505015631  454.591352064
```

The model is quite acceptable, but we can improve the diagnosis through the commands of the *car* package.

The following syntax is used to obtain the Durbin Watson statistic:

```
> library("car", lib.loc="C:/Users/usuario/Documents/R/win-li
brary/3.2")
> durbinWatsonTest(reg1)
 lag Autocorrelation D-W Statistic p-value
 1         0.3320878        1.335684       0
 Alternative hypothesis: rho ! = 0
```

It is observed that the value 1.33 of the D-W statistic allows suspecting the presence of autocorrelation, certified by the p-value zero of the formal contrast of Durbin Watson.

To analyze the heteroscedasticity of the model we use the following syntax:

```
> ncvTest(reg1)
Non-constant Variance Score Test
Variance formula: ~ fitted.values
Chisquare = 1372.616     Df = 1        p = 1.875088e-300
```

The low p-value contrast indicates that there is no heteroscedasticity in the model.

The modulus of the standardized residues against the adjusted values is random (Figure 7-34), which confirms the absence of heteroscedasticity problems.

```
> spreadLevelPlot(reg1)
```

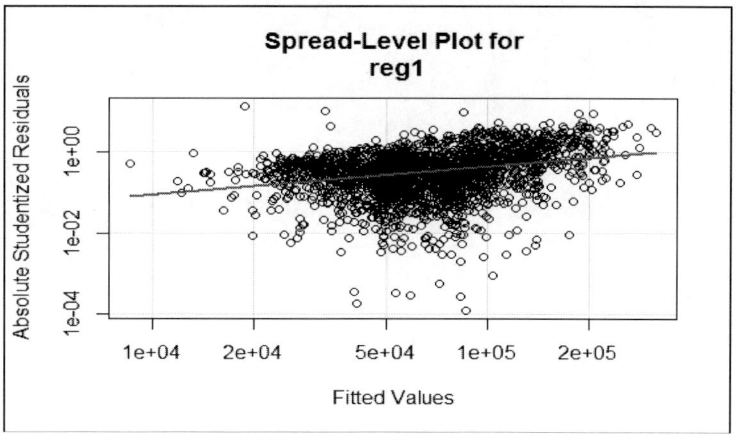

Figure 7-34

Then we made the graphs of the independent variables against the dependent variable (Figure 7-35), observing a good linear fit, which certifies that there are no problems of nonlinearity.

```
> windows()
> avPlots(reg1)
```

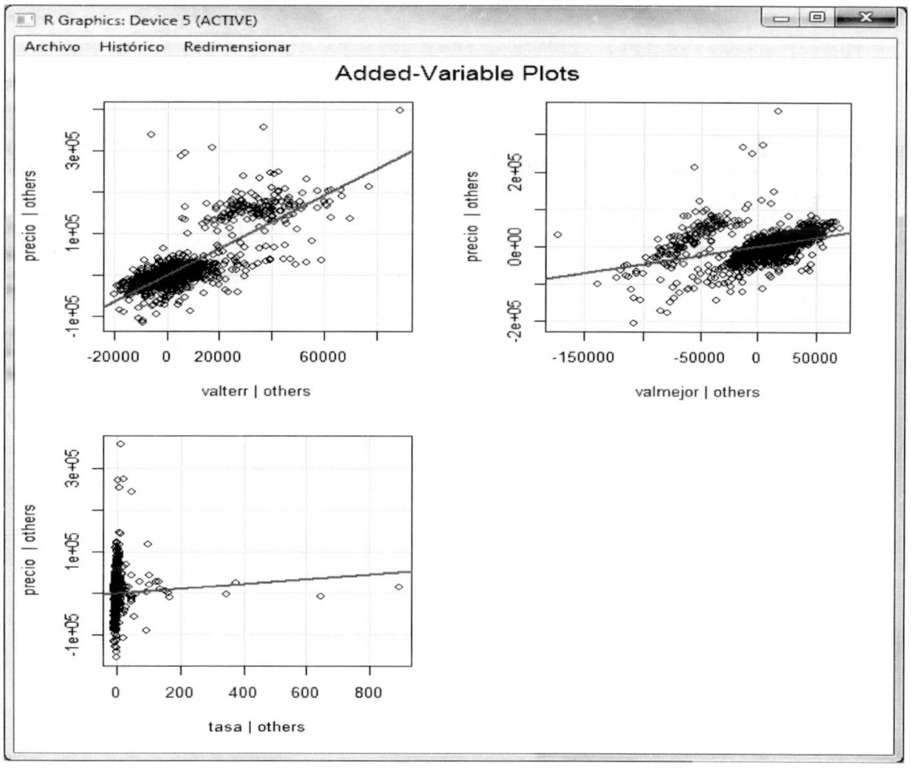

Figure 7-35

The graph of independent variables against compinents plus residues (Figure 7-36) leads us to the same conclusion about linearity.

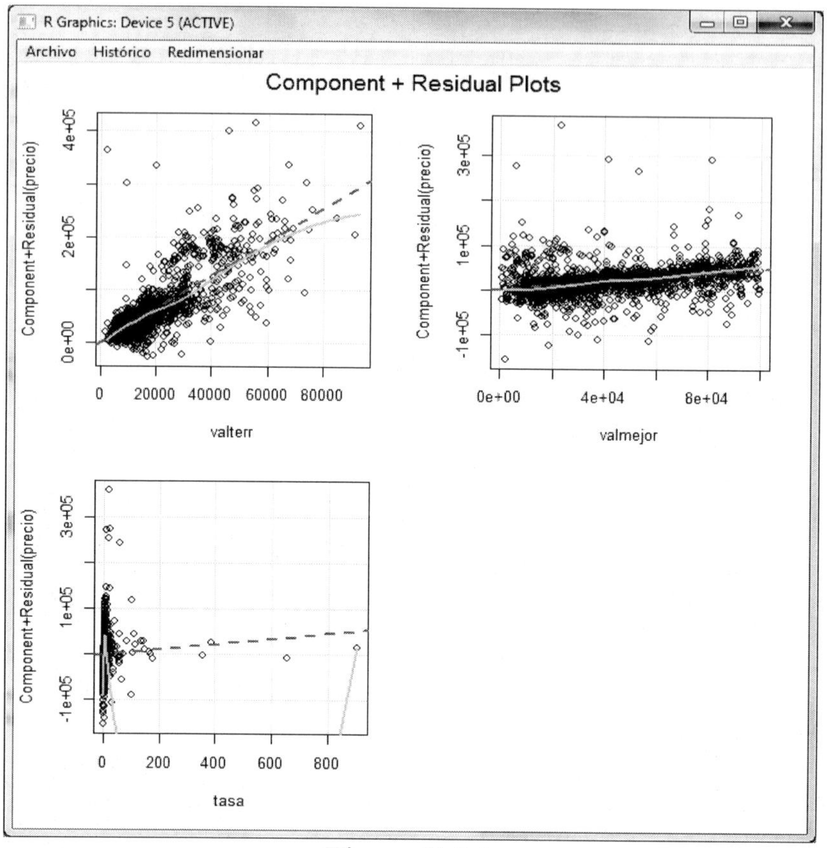

Figure 7-36

Exercise 7-2. Using the file viviendas.sav that contains information about neighborhoods in a city, price of the houses, price of the land, value of the improvements and ratio between real price of the house and appraised price, estimate a multinomial logit model that explains the probability that a house belongs to a certain neighborhood depending on the price of the house, the price of the land, the value of the improvements and the ratio between real price and appraised price

We start by loading n memory the *nnet* packet and using the *multinom* command syntax as shown in the figure:

```
> library("nnet", lib.loc="C:/Program Files/R/R-3.2.0/library")

> reg2= multinom(neighborhood ~ price + valterr + valmejor +
rate, contrasts=TRUE)

# weights: 42 (30 variable)
initial value 4748.020764
iter 10 value 4333.254672
```

```
iter 20 value 4319.164119
iter 30 value 3455.721475
iter 40 value 3070.023121
iter 50 value 3064.738076
iter 60 value 3064.651888
iter 70 value 3064.632879
iter 80 value 3064.630572
iter 80 value 3064.630553
iter 80 value 3064.630526
final value 3064.630526
converged

> summary(reg2)

Call:
multinom(formula = neighborhood ~ price + valterr + valmejor
+ tax,
     contrasts = TRUE)

Coefficients:
 (Intercept) valterr              price          valmejor          ra
te
B      4.121860 -4.186222e-06 -4.714150e-05 3.093752e-06 0.006161963
C      5.371426 -1.855751e-05 -2.059026e-05 -6.985867e-06 0.006226186
D      5.583556 -3.490294e-05 -7.743415e-05 4.545140e-05 0.005327313
E      9.437751 -3.560329e-05 -1.202013e-04 -1.345121e-05 -0.05335403
4
F     11.755051 -4.928927e-05 -5.745320e-05 -9.749678e-05 -0.02852782
5
G     16.269462 -6.220735e-05 -3.368239e-04 -1.296163e-04 0.013898775

Std. Errors:
    (Intercept) valterr              price          valmejor          ra
te
B 6.341515e-07 2.851473e-06 1.220322e-05 4.812793e-06 1.200368e-03
C 1.024628e-06 3.419921e-06 1.353138e-05 4.998696e-06 1.389694e-03
D 6.829220e-07 4.904475e-06 1.594236e-05 7.050610e-06 1.071451e-03
E 6.274641e-08 4.759482e-06 1.778867e-05 6.506805e-06 3.972753e-07
F 4.670940e-08 4.444455e-06 1.708155e-05 6.170850e-06 6.553111e-08
G 2,166476e-08 6,633832e-06 2,348407e-05 9,525667e-06 1,591271e-05

Residual Deviance: 6129,261
AIC: 6189.261
```

A good estimate with very small standard errors is observed.

You could also have used the multinom command from the mlogit library, using the following syntax.

```
> library("mlogit", lib.loc="C:/Users/user/Documents/R/win-li
brary/3.2")

> reg3= multinom(neighborhood ~ price + valterr + valmejor +
tax)
# weights: 42 (30 variable)
initial value 4748.020764
iter 10 value 4333.254672
iter 20 value 4319.164119
iter 30 value 3455.721475
iter 40 value 3070.023121
iter 50 value 3064.738076
iter 60 value 3064.651888
iter 70 value 3064.632879
iter 80 value 3064.630572
iter 80 value 3064.630553
iter 80 value 3064.630526
final value 3064.630526
converged
> summary(reg3)
Call:
multinom(formula = neighborhood ~ price + valterr + valmejor
+ tax)

Coefficients:
 (Intercept) valterr                  price         valmejor          ra
te
B     4.121860 -4.186222e-06 -4.714150e-05 3.093752e-06 0.006161963
C     5.371426 -1.855751e-05 -2.059026e-05 -6.985867e-06 0.006226186
D     5.583556 -3.490294e-05 -7.743415e-05 4.545140e-05 0.005327313
E     9.437751 -3.560329e-05 -1.202013e-04 -1.345121e-05 -0.05335403
4
F    11.755051 -4.928927e-05 -5.745320e-05 -9.749678e-05 -0.02852782
5
G    16.269462 -6.220735e-05 -3.368239e-04 -1.296163e-04 0.013898775

Std. Errors:
    (Intercept) valterr                  price         valmejor          ra
te
B 6.341515e-07 2.851473e-06 1.220322e-05 4.812793e-06 1.200368e-03
C 1.024628e-06 3.419921e-06 1.353138e-05 4.998696e-06 1.389694e-03
D 6.829220e-07 4.904475e-06 1.594236e-05 7.050610e-06 1.071451e-03
E 6.274641e-08 4.759482e-06 1.778867e-05 6.506805e-06 3.972753e-07
F 4.670940e-08 4.444455e-06 1.708155e-05 6.170850e-06 6.553111e-08
G 2,166476e-08 6,633832e-06 2,348407e-05 9,525667e-06 1,591271e-05

Residual Deviance: 6129,261
AIC: 6189.261
```

The result is the same.

Exercise 7-3. Solve the previous exercise by means of an ordered multinomial logit model.

We will start by loading the MASS library in memory and then use the command *polr*.

```
> library("MASS", lib.loc="C:/Program Files/R/R-3.2.0/library
")

> reg4= polr(neighborhood ~ price + valterr + valmejor + rate
, method="logistic")

> reg4
Call:
polr(formula = neighborhood ~ price + valterr + valmejor + ra
te, method = "logistic")

Coefficients:
        valterr         price          valmejor             rate
-2.384657e-05 -5.416448e-05 -3.714585e-05 1.727940e-03

Intercepts:
       A|B         B|C          C|D          D|E          E|F          F|G
-10.621977 -7.170973 -6.098913 -4.591116 -3.061032 -1.840495

Residual Deviance: 6917.487
AIC: 6937.487
```

Exercise 7-4. The logitb.sav file containing data from a sample of 53 patients with prostate cancer is considered, measuring age, the acid level that measures the extent of the tumor, the degree of aggressiveness of the tumor, the stage of the tumor, the results of an X-ray and when it was detected during surgery that the cancer had spread to the lymph nodes. Based on this data, the idea is to adjust a model that makes it possible to predict when the cancer has spread to the lymph nodes (or not) without the need for surgical intervention.

We begin by reading the logitb.sav file, observing its content and enabling its variables.

```
> library(foreign)
> datosc=read.spss("H:/R/data/logitb.sav", use.value.labels=T
RUE)

> summary(datosc)
          Length Class Mode
ID        53      -none- numeric
```

```
RADIOGRA 53        -none- numeric
STAGE     53       -none- numeric
GRADE     53       -none- numeric
AGE      53        -none- numeric
ACID     53        -none- numeric
NODES     53     numeric factor
```

```
> attach(datosc)
```

Then we adjust the logit model ordered.

```
> logistical=glm(NODOS~RADIOGRA+STAGE+DEGREE+AGE+ACID, family
=binomial(link="logit"))
> summary(logistical)

Call:
glm(formula = NODES ~ RADIOGRA + STAGE + GRADE + AGE + ACID,
 family = binomial(link = "logit"))

Deviance Residuals:
 Min         1Q     Median         3Q        Max
-2.0110 -0.7020 -0.3654    0.5723    1.9852

Coefficients:
             Estimate Std. Error z value Pr(>|z|)
(Intercept)   0.06180    3.45992    0.018    0.9857
X-RAY         2.04534    0.80718    2.534    0.0113 *
STAGE         1.56410    0.77401    2.021    0.0433 *
GRADE         0.76142    0.77077    0.988    0.3232
AGE          -0.06926    0.05788   -1.197    0.2314
ACID          0.02434    0.01316    1.850    0.0643 .
---
Meaning codes: 0 '***' 0.001 '**' 0.01 '*' 0.05 '.' 0.1 ' ' 1

(Dispersion parameter for binomial family taken to be 1)

    Null deviance: 70.252 on 52 degrees of freedom
Residual deviance: 48.126 on 47 degrees of freedom
AIC: 60,126

Number of Fisher Scoring iterations: 5
```

It is observed that the constant is not significant. We will then adjust the model without constant.

```
> logistical=glm(NODOS~0+RADIOGRA+STAGE+GRADE+AGE+ACID, famil
y=binomial(link="logit"))
> summary(logistical)
```

```
Call:
glm(formula = NODES ~ 0 + RADIO + STAGE + GRADE + AGE + ACID,
 family = binomial(link = "logit"))
Deviance Residuals:
 Min         1Q     Median        3Q        Max
-2.0113 -0.7029 -0.3665    0.5727    1.9872

Coefficients:
          Estimate Std. Error z value Pr(>|z|)
X-RAY      2.04560     0.80711    2.534 0.01126 *
STAGE      1.56566     0.76919    2.035 0.04180 *
GRADE      0.76284     0.76648    0.995 0.31961
AGE       -0.06830     0.02131   -3.205 0.00135 **
ACID       0.02439     0.01294    1.884 0.05955 .
---
Meaning codes: 0 '***' 0.001 '**' 0.01 '*' 0.05 '.' 0.1 ' ' 1

(Dispersion parameter for binomial family taken to be 1)

     Null deviance: 73.474 on 53 degrees of freedom
Residual deviance: 48.126 on 48 degrees of freedom
AIC: 58,126

Number of Fisher Scoring iterations: 5
```

The model is already more or less acceptable, except perhaps the GRADE variable whose significance is 70%.

Exercise 7-5. Solve the previous exercise using a Probit model.

We begin by estimating the model through probit regression.

```
> probit1=glm(NODES~RADIOGRAPH+TAPE+DEGREE+AGE, family=binomi
al(link="probit"))
> summary(probit1)

Call:
glm(formula = NODES ~ RADIOGRA + STAGE + GRADE + AGE + ACID,
 family = binomial(link = "probit"))

Deviance Residuals:
 Min         1Q     Median        3Q        Max
-2.0061 -0.7001 -0.3230    0.5813    1.9760

Coefficients:
```

```
              Estimate Std. Error z value Pr(>|z|)
(Intercept)   0.065353   2.036217   0.032 0.97440
RADIOGRA      1.211738   0.461926   2.623 0.00871 **
STAGE         0.955897   0.443673   2.155 0.03120 *
GRADE         0.425286   0.449035   0.947 0.34358
AGE          -0.042297   0.033914  -1.247 0.21233
ACID          0.015021   0.007759   1.936 0.05289 .
---
Meaning codes: 0 '***' 0.001 '**' 0.01 '*' 0.05 '.' 0.1 ' ' 1

(Dispersion parameter for binomial family taken to be 1)

     Null deviance: 70.252 on 52 degrees of freedom
Residual deviance: 47.828 on 47 degrees of freedom
AIC: 59,828

Number of Fisher Scoring iterations: 6
```

It is observed that the constant is not significant, which leads us to estimate the model without a constant.

```
> probit1=glm(NODES~0+RADIOGRAPH+TAPE+GRADE+AGE+ACID, family=
binomial(link="probit"))
>
>
> summary(probit1)

Call:
glm(formula = NODES ~ 0 + RADIO + STAGE + GRADE + AGE + ACID,
 family = binomial(link = "probit"))

Deviance Residuals:
 Min       1Q    Median       3Q       Max
-2.0066 -0.7017 -0.3255    0.5822    1.9795

Coefficients:
           Estimate Std. Error z value Pr(>|z|)
RADIOGRA   1.211718   0.461858   2.624 0.008701 **
STAGE      0.956375   0.442234   2.163 0.030572 *
GRADE      0.427660   0.446311   0.958 0.337956
AGE       -0.041276   0.011823  -3.491 0.000481 ***
ACID       0.015067   0.007621   1.977 0.048034 *
---
Meaning codes: 0 '***' 0.001 '**' 0.01 '*' 0.05 '.' 0.1 ' ' 1
(Dispersion parameter for binomial family taken to be 1)
```

```
    Null deviance: 73.474 on 53 degrees of freedom
Residual deviance: 47.829 on 48 degrees of freedom
AIC: 57,829

Number of Fisher Scoring iterations: 5
```

The model is already more or less acceptable, except perhaps the GRADE variable whose significance is 70%. The results are very similar to the case of the logit model.

Made in United States
North Haven, CT
06 November 2026

21817329R00139